Adapting Science Fiction to Television

Science Fiction Television
Series Editor: A. Bowdoin Van Riper

From Starship Captains to Galactic Rebels: Leaders in Science Fiction Television, by Kimberly Yost, 2014

Joss Whedon's Dollhouse*: Confounding Purpose, Confusing Identity*, edited by Sherry Ginn, Alyson R. Buckman, and Heather M. Porter, 2014

Doctor Who *and the Art of Adaptation: Fifty Years of Storytelling*, by Marcus K. Harmes, 2014

The Language of Doctor Who*: From Shakespeare to Alien Tongues*, edited by Jason Barr and Camille D. G. Mustachio, 2014

Firefly *Revisited: Essays on Joss Whedon's Classic Series*, edited by Michael Goodrum and Philip Smith, 2015

The Paranormal and the Paranoid: Conspiratorial Science Fiction Television, by Aaron John Gulyas, 2015.

Adapting Science Fiction to Television: Small Screen, Expanded Universe, by Max Sexton and Malcolm Cook, 2015.

Adapting Science Fiction to Television

Small Screen, Expanded Universe

Max Sexton
Malcolm Cook

ROWMAN & LITTLEFIELD
Lanham • Boulder • New York • London

Published by Rowman & Littlefield
A wholly owned subsidiary of The Rowman & Littlefield Publishing Group, Inc.
4501 Forbes Boulevard, Suite 200, Lanham, Maryland 20706
www.rowman.com

Unit A, Whitacre Mews, 26-34 Stannary Street, London SE11 4AB

British Library Cataloguing in Publication Information Available

Library of Congress Cataloging-in-Publication Data

Sexton, Max, 1966–
Adapting science fiction to television : small screen, expanded universe / Max Sexton and Malcolm Cook.
pages cm. — (Science fiction television)
Includes bibliographical references and index.
ISBN 978-1-4422-5269-1 (hardback : alk. paper) — ISBN 978-1-4422-5270-7 (ebook) 1. Television adaptations—History and criticism. 2. Science fiction television programs—History and criticism. 3. Science fiction—Television adaptations. I. Cook, Malcolm, 1976– II. Title.
PN1992.8.T45S39 2015
791.45'6—dc23
2015007647

™ The paper used in this publication meets the minimum requirements of American National Standard for Information Sciences Permanence of Paper for Printed Library Materials, ANSI/NISO Z39.48-1992.

Printed in the United States of America

To Alison Smith and Gino Marcangelo
—MS

To Mum and Dad
—MC

Contents

Introduction

Adapting Science Fiction to Television

On initial inspection the title of this book would seem to offer a self-evident methodology and field of study, especially when addressing a limited period between the 1950s and 1980s and conducted using case studies, as is the situation here. As discussion throughout this book attests, all the terms in the title of this book share a linked history. Television has been held in low cultural esteem due to its association with mass entertainment. The process of adaptation has been used as a device to raise its status. In some cases the relay function of television implied it was a transparent medium that introduced no transformation and offered strict fidelity to the original source, as in the presentation of theatre. In other cases adaptation was conceived as a more fluid and active process that allowed television as having some form of intrinsic nature, but nevertheless favored literary sources that provided the maximum cultural value and established a hierarchy between art forms. While science fiction is a genre that appears in a range of arts and its valuation has consequently been uneven, like television on which it was an important presence, it has commonly been held in low cultural esteem and associated with particular generic characteristics. The result of this linked history has been that all three terms in our title have attracted dominant meanings in general usage, especially within the industry, that have also restricted and directed scholarly investigation of them.

This book will propose that the uncritical use of these terms masks a historical and cultural construction that we aim to uncover and examine through our case studies. In doing so it will reexamine key questions the terms conceal: Does television have an intrinsic nature? Is science fiction a fixed label? How was adaptation first conceived? Given this, before addressing the television programs that are the primary subjects of this book, it is necessary to examine the usage of these terms within the academic discourse of adaptation, especially in regard to television, and clarify the new understanding this book brings to them.

TELEVISION ADAPTATION AND THE MYTH OF MEDIUM SPECIFICITY

Adaptation as a concept is central to television in a number of ways. As several of the case studies in this book will show, the adaptation of stories, characters, and aesthetic practices from other forms has been familiar since its earliest days, and such originating material has been marshaled together without necessarily exploiting a perceived intrinsic nature of television. The academic study of adaptation is commonly traced back to George Bluestone's 1957 *Novels into Film*, and forms a substantial body of scholarly work that may be classified as "adaptation studies."[1] Yet this work remains restricted in a number of ways. First, it is heavily invested in literature and literary studies both as a subject for study and methodologically. As Thomas Leitch has argued, there is a tendency to structure such work around canonical texts and authors, and doing so "makes fidelity to the source text central to the field."[2] Even in those cases where such assumptions are questioned, literature remains a central point of reference and literary studies provide the primary methodology for the analysis of "texts." The second limitation of adaptation studies is in their handling of moving images, in which they reproduce the long-standing subordination of television in relation to cinema, although notable exceptions to this dominant trend include Sarah Cardwell's work on television and the classic novel and J. P. Telotte and Gerald Duchovnay's edited collection on Science Fiction adaptation for film and television, which will be discussed later in this introduction.[3]

This book is concerned with a wider understanding of adaptation in three ways. First, while the adaptation of literary texts is considered, these are not given a priori promotion and a wide range of other sources are considered, including cinema, radio, and comics. Furthermore, no hierarchy is established between these forms, and the notion of "fidelity" is interrogated rather than taken for granted. The second expansion of adaptation undertaken here is to consider not simply the relationship between any individual source and its televisual adaptation, but also the adaptation of aesthetic qualities between media more generally, especially the way television selectively adopted and adapted characteristics commonly seen as typical of cinema, and moreover seen as antithetical to television. A precedent for such an approach can be found in J. P. Telotte's discussion of *The Twilight Zone*, a show that was predominantly written directly for television rather than adapted. Despite this, Telotte makes a persuasive case for understanding this show as adapting cinematic qualities for television.[4] Finally, arising from the prior point, the very concept of adaptation raises the question of medium specificity, of the idea that different media have inherent qualities that differentiate them and constitute the basis of them as distinct art forms, and may even be used as aesthetic criteria for their valuation and judgment.

The frequent occurrence of the generation of works of art in different media sharing a source clearly suggests the possibility of common characteristics or qualities that exist at a transmedial level. Nevertheless, the term *adaptation* also inherently posits distinctions between the media being referred to that require the process of adaptation to bridge them.[5] As such adaptation would seem to uphold the concept of medium specificity, of different media having essential qualities that distinguish them.

Medium specificity remains a dominant argument for creating boundaries and defining media, especially for what Noel Carroll calls "self-consciously invented arts" such as photography, film, and video.[6] Writers engaging with these relatively new art forms that are dependent upon recent technological developments have seen the need to define and legitimize them in comparison to both long-established art forms as well as each other. Thus film was commonly distinguished from both theatre and photography, while television was defined in contrast to film and radio. In the 1980s, Carroll explored the various arguments used in making claims for medium specificity and convincingly argued for their lack of philosophical coherence.[7]

Present-day debates about digital technologies and the effect they have upon media have further undermined such arguments in relation to moving images. Digital technologies have introduced simultaneously a convergence and splintering of media. On the one hand the elimination of film as a dominant transcription format has left all moving images being captured by the same technologies, eliminating a key differentiating quality between television and cinema, one that will be discussed and questioned throughout this book. On the other hand, a proliferation of exhibition technologies has unsettled long-standing assumptions about the distinctions between cinema and television. Rather than a big screen/small screen distinction, we now have a continuum of screen sizes, from the largest theatrical presentations to large high-definition domestic displays and mobile devices in a range of sizes. Likewise home video formats and live broadcasting into cinemas undermine simple distinctions between transmission and transcription media. While these changes have become pronounced more recently, their implications for the medium-specificity thesis were observed by Carroll over a decade ago.[8]

Some writers may look to define "digital" as a new dominant medium of its own, with new medium-specific qualities, yet this would simply replicate the flaws Carroll elucidates. Another approach is offered in a recent discussion of the relationship between television and other arts by Sarah Cardwell. In it she draws valuable attention to distinctions between the use of "media/ medium" to describe base materials and the use of these terms to describe art forms, encompassing both their technology and practices. In doing so she acknowledges the imprecision of the notion of a "medium," and hopes to clarify it by contrasting it with "art forms" as an identifier, removing the dependence upon specific technological base media, and thereby hoping to evolve greater precision going forward.[9]

This book, however, is concerned with a historical period where the imprecision in the discussion of "media" that Cardwell hopes to eliminate was pervasive and played a formative role in how television developed, and thus it cannot simply be ignored. In another discussion of recent changes in technology William Uricchio uses the present "crisis" in film studies to review a range of examples from film and television history that complicate and blur the boundaries between traditional medium-specific forms. In doing so Uricchio demonstrates that cinema and television have been overdetermined, and argues that, rather than adopt ahistoric definitions, these forms should be understood as historically contingent and studied in these terms.[10]

This book presents one such approach to the history of television and offers an account of the complex relationship between television and other media that provides evidence for the commingling of media and thus upholds Uricchio and Carroll's arguments that medium specificity is a flawed tool for the study of moving images. Yet medium specificity remains central to this study, not as an ahistorical ontology, but rather as a discourse that was in general circulation for producers, commentators, and viewers and as such shaped the development of the programs discussed. The process of adaptation especially brings this to the fore, with filmmakers engaging and negotiating the specificity of television in relation to the sources being used and critics and audiences judging work on the basis of their definitions of media. Rather than simply eliminating difference between media this book acknowledges the contingent nature of medium-specific definitions. In so doing it adopts the case study as the best model for close examination of historically and culturally specific developments in comparative definitions of media. There can be no history of television if there is no singular definition of it as a medium, but there can be a history of the way that medium was defined at particular times and in particular places, and this may prove instructive for present-day debates about the future of media and their study.

GENRE AND TELEVISION

The treatment of medium specificity described above has close parallels with the treatment of genre in film and television studies. While genres have in the past been understood as fixed, ahistoric structures, they have also been long understood as being composed of historically and culturally shifting sets of traits and expectations.[11] Science fiction in the twentieth century is an exemplar of the changing construction of genre definitions, not least due to the fundamental shifts in the wider social context. The clearest example of this is the space race and the exploration of the moon in the 1960s, events which had been anticipated by earlier science fiction work, but could hardly have been considered immediately plausible at the turn of the century. Other examples are also pertinent, such as the development of nuclear energy and associated technologies. These technologies allowed the launch in 1954 of the American nuclear-powered submarine *Nautilus*, its name a reference to Jules Verne's *Twenty Thousand Leagues under the Sea* (1869), which is discussed in more detail in chapter four.[12] Here was an event Verne had imagined as a voyage extraordinaire, which would be reimagined as "science fiction" on film and television in the twentieth century and then served as inspiration for scientific and technological fact in the 1950s.

Science fiction as a popular form offers two characteristics that make it particularly valuable for the study of adaptation and the examination of the construction of medium specificity. First, it has been exploited and adapted in all the different media examined here, including radio, comic books, television, and cinema. This offers an opportunity to consider the diversity of possible sources for adaptation to television, with a resultant diversity of forms, in contradistinction to the restricted view offered by the consideration of exclusively literary sources. Secondly, despite such apparent diversity, science fiction has also been seen as having affinity with particular modes of production and tied into institutional debates about medium specificity. In terms of moving images, influential accounts such as Annette Kuhn's arguments about the "mobilization of the visible" and Susan Sontag's description of the genre's "imagery of destruction" align it with spectacle and cinema.[13] While there is certainly a tradition of science fiction filmmaking that is predicated on a shared investment in spectacle, this book aims to show that the adaptation of science fiction for television complicates notions of spectacle and narrative as much as the complex relations between genre and the medium. As will be shown, spectacle has played a role in television science fiction, but science fiction on television has explored a number of different modes and forms. Such multiplicity reflects a commitment to mass entertainment, as well as catering to a specific audience interested in science fiction. Consequently, analysis that reflects how genres actually form and evolve rather than the overreliance on notions of television as a popular medium, as a relay device, or the use of any general paradigm avoids the need for claims about a generalized mode on television.

An examination of television's ability to create a taxonomy encompassing some of the main strands of the science fiction genre begins by establishing the original intrinsic nature of the medium before dealing with questions of institutional discourses and practices specific to particular television programs.

As Noel Carroll suggests, from the 1950s "theoretical spokespersons for the new medium were at pains to differentiate TV from film—to show that TV was not just an ersatz form of cinema."[14] The possibility of television developing as an aesthetic medium and offering a diverse set of programming further complicates claims about intrinsic qualities of the medium. It is impossible therefore to make a generalization about the programs that are analyzed in this book, due to their often variegated nature, except that they are all "science fiction." But a greater use of visual style overall did allow for the creation of distinctive forms of programs both in the United States and UK.

In the initial period under observation, television did not use genre to the same extent as the cinema because it had a property of "flow" that blended programs together to form an overall evening's worth of programming. Many critics, such as Raymond Williams and John Ellis, both of whom are considered in this book, have argued that the unit of coherence for television is at a level larger than the program and different from the genre, although the book examines what contribution a genre, with its fixed codes of narrative and visual style, could make to flow. Crucially, this book demonstrates how the recording of a program on film in the 1950s gave it a fixity and a nonephemeral quality that much of television lacked as it continued to use the tropes of liveness. Later, the adaptation of science fiction on television continued to guarantee meanings and pleasures for audiences, even if it became, by the mid-1960s, an unstable marker of expectation and knowledge in programs such as *Out of the Unknown*.

As a consequence, the study of genre on television differs in many ways from the cinema. By asking how the programs in the book were understood by institutions, program makers, as well as audiences, the text can be analyzed as an industrial form of aesthetic practice. The analysis of genre goes beyond the limits of the text and examines a number of "extra-textual" areas including the production methods and personalities within television at particular historical moments. How science fiction was first viewed and understood on television serves two purposes: First, it pays careful attention to the historical conditions that formulated and reformulated the science fiction series and/or anthology on television, and how it was understood by the institution at the time. Second, it examines what was meant by "creativity" and/or authorship at a time when television was both technologically and institutionally constrained.

MODE AND THE EXPANDED FUNCTION OF TELEVISION

The critical discourse constructed around the early function of television was concerned with television as an aesthetic medium, but which aesthetic was "natural" to television? In 1970, Joan Bakewell was to comment that, "a whole theology was erected to justify television's existence as an autonomous art in terms of its liveness."[15] If such debates were raised it was television's artistic and ontological specificity, its relations to other media, particularly cinema and radio, and its proper program forms, that were core to the debate.

It can be argued that the lack of recorded drama programs that could be stored and later reviewed, and improved on, or reinterpreted, essentialized television into a perpetual present during the 1950s and 1960s, reinforcing its earlier relay mode. The problems of recording and an adherence to the simple relay function of television meant that "The vast majority of the single plays and serials produced by the BBC until the 1960s were adaptations, coming to the viewer with a prior seal of approval from the West End theatre. . . . Plays written specifically for television were the exception rather than the rule."[16]

Direct transmission remained the dominant mode, but it was possible to cut cleanly between studio cameras as an alternative to mix or fade out/fade-in. For Charles Barr, this "brings the fundamental syntax of television a bit closer to that of cinema."[17] The organization of space was becoming closer to mainstream cinema and the multicamera studio drama can be understood as a basic version of the Hollywood continuity system. On the other hand, the images shot live by television cameras may have looked superficially like an edited film sequence, but they were a record of real-time continuity and not, as in the case of film, the construction of an apparent continuity. One advantage of this was that instead of using a single film camera and the time-consuming business of organizing and shooting separate setups, multicamera studio television could be shot continuously.

This then was a fundamental difference between shooting on film and broadcasting live within the electronic studio. However, the use of film and later the recording on video continued to offer the possibility of allowing for the manipulation of time across cuts and avoid the strict naturalistic/theatrical continuities of space and time.

By the mid 1950s, television was no longer following the logic of remaining a transmitting medium, but was becoming one that, like cinema before it, could produce images that might manipulate visual information using separate film camera setups. Nigel Kneale, the writer of the *Quatermass* series, explained that "Film adds both physical freedom and atmosphere. . . . All the technically difficult scenes, involving special effects which it would have been risky to tackle live, were filmed, giving the producer much greater control."[18]

By experimenting with new visually expressive techniques, writers and directors sought to always try to provide something worth watching for the peak time *majority* audience, as well as the audience interested in science fiction. In this way, a diverse dramatic form might be created that was entertaining and relied on earlier regulated forms, yet could remain serious minded and depend on ideas to do with authorship and "creativity": a liminal state between an ideas-led science fiction genre and television's culture of mass entertainment. This is especially evident in *Science Fiction Theatre* in the United States and even more so in *Quatermass and the Pit* in the UK.

This book hopes to demonstrate that issues to do with the adaptation of science fiction for television cannot be explained in reference to one logic: textual, institutional, or technological. By focusing on the dynamics of some of the key players producing science fiction for television, it is possible to determine shifting values and understand the interconnectedness of the television text, the science fiction genre *and* the economic/technical dimensions of the television series.

In these ways, the adaptation of science fiction as television drama deployed action and spectacle that could exploit generic verisimilitude and authorship in ways prescient of modern-day television programs. Rather than suggesting a dominant paradigm for television, the process of adaptation reveals how personal stylistic decisions to do with narrative and style can often be understood as accommodations to production conditions. For example, a director such as Rudolph Cartier working within the BBC could exercise considerable control in the production process of *Quatermass and the Pit*. The lighting used in the program reveals that if television is a mass medium, it is also a medium capable of including personal predilections in some of its content. This book seeks to demonstrate how the adaptation of science fiction on television should be regarded as an assemblage of several stylistic variables.

The purpose of the case studies therefore is to identify where and when such possibilities existed and their operation and inscription in the programs that have been selected. Some of the main production discourses of the time have been described in order to make comparisons between dominant notions about television drama and the efforts by writers, directors, and producers at the time to modify, as well as counter these notions. The book seeks to establish how expectations about science fiction both within and outside television allowed for a complex exchange of forms and modes to function as an alternative to the fidelity of its sources. Each case study seeks to complicate any understanding of television as a medium that privileges the spoken word over the image or that offers a continuous flow of programs glanced at by a distracted viewer. It seeks to retheorize such paradigms by offering an alternative based on not simply formal or aesthetic approaches to texts, but how systems of cultural hierarchies to do with film and video, popular and serious drama, came to be understood and managed.

Chapter 1 examines how in its formative period television was predominantly live. However, it discusses the tension between television's liveness and the use of a recorded performance on film. The chapter also introduces the reader to notions of how television borrowed conventions from radio and Hollywood, to form its own system of signification and a mode of narrative suited to the small screen that was addressing an audience watching a domestic medium. It examines how the first sci-fi television series on US television reworked the formulas and types of narrative popular from the cinema chapter-play or the so-called "fantastic cliff-hangers." Other types of programs such the television anthology, *Science Fiction Theatre*, are examined to offer fresh insights into the problems of adapting a program to a televisual form that had to reconcile (cinematic) spectacle and the sense of immediacy on the small screen.

Chapter 2 examines how in the UK the BBC brought a new sophistication and cultural value to the adaptation of science fiction on television. The success of the *Quatermass* science fiction series marked a moment of original, as distinct from preexisting drama, and signaled an expansion of the function of television as a representation of the fantastic. The chapter examines how the final series, *Quatermass and the Pit* (1959), adapted ambitious production values and spectacle without the loss of its important small-screen cultural values, championed by the British Broadcasting Corporation. In this way, the source material, Nigel Kneale's script, was adapted to a filmic mode and also formed an attempt to modify an earlier mode of television drama.

Chapter 3 examines the BBC science fiction show *Out of the Unknown*, an anthology drama produced during the 1960s, and the dramatization of the science fiction short story. Most of the episodes, produced by Irene Shubik, were based on a set of values from the liberal arts that strove to guide the audience's tastes and their cultural development, but they also deployed a mix of different styles in order to provide a form of mass entertainment.

Chapter 4 explores the diversity of approaches to adapting animated science fiction for television within a three-decade period. The earliest use of animation on television was the adaptation of theatrical animated cartoons, which established a new definition for animation and its audience within a television context as well as demonstrated a new market for science fiction characters. This development, in turn, prompted the production of made-for-television animation, including *The New Adventures of Superman*. A study of this show considers it within the context of multiple earlier adaptations and considers the way the Superman mythos and the duality of the character were used to negotiate and redefine medium specificities. In the wake of broader social events in the late 1960s, aspects of this new definition were subject to scrutiny and criticism by many commentators. As a result animation was again redefined, with the adaptation of canonical literary sources offering protection from the criticisms aimed at comic book adaptations. The adaptation of Jules Verne's *Twenty Thousand Leagues under the Sea* is offered as a case study.

Finally, chapter 5 compares the two BBC-produced *Day of the Triffids* programs in 1981 and 2009 as examples of the growing plurality of television in its approach to adaption. Both series adapted the original novel by John Wyndham, written in the 1950s. However, since the 1980s, the growth of digital computer special effects demonstrates how far the television series has become separated from earlier generic forms, and has accommodated a form that raises new questions about the relationship between spectacle, action, and narrative. The increased semiotic density in the 2009 adaptation of *The Day of the Triffids,* unlike the minimalist style of the 1981 adaptation, suggests how film and television convergence and the development of a heterogeneous televisual style may be understood.

The question of television adaptation is not one that is best understood as a single process that relies on a single paradigm. Such responses to television have set limits that have never been ontologically sound. Instead, the evolution and versatility of television is central to the debate being raised here in this volume, which seeks to demonstrate less of a technological difference between television and older media, and more a philosophical difference in its adaptation of preexisting material. Such differences rely on a firm historical understanding of both the development of production tendencies in television to account for changes from a live to a recorded mode and also the growing function of television as it sought to offer more sophisticated forms of dramatic narrative within the genre of science fiction.

NOTES

1. Bluestone, *Novels into Film.*
2. Leitch, *Film Adaptation and Its Discontents*, 3.
3. Cardwell, *Adaptation Revisited*; Telotte and Duchovnay, *Science Fiction Film, Television, and Adaptation.*
4. Telotte, "The Cinematic Zone of *The Twilight Zone*."
5. This may be brought into focus by considering the implications of terms other than *adaptation* used to describe similar activities in a variety of media: *translation*, *remake*, *cover version*, *production*, *performance*.
6. Carroll, "Medium Specificity Arguments."
7. Carroll, "The Specificity of Media in the Arts"; Carroll, "Medium Specificity Arguments." These arguments were later collected in Carroll, *Theorizing the Moving Image.*
8. Carroll, *Engaging the Moving Image*, 265–80.
9. Cardwell, "Television amongst Friends."
10. Uricchio, "Film, Cinema, Television . . . Media?," 266.
11. Schatz, *Hollywood Film Genre as Ritual.*
12. *Times*, February 2, 1954, 7.
13. Kuhn, *Alien Zone*, 6; Sontag, *Against Interpretation*, 213.
14. Carroll, *Engaging the Moving Image*, 265–66.
15. Bakewell and Garnham, *The New Priesthood*, 14.
16. Caughie, *Television Drama*, 38.
17. Barr, "They Think It's All Over," 58.
18. Kneale, "Not Quite So Intimate," 87.

Chapter One

Primeval Dawn

"Man in Space" and Science Fiction Theatre

> Television, the child of three parents—theatre, film, and radio—found it hard to establish its own identity in its early days. While television experimented with its potential and strived to determine its own creative identity, practitioners from its parent arts gave it unkind names. Some theatre actors who had fared poorly before the electronic cameras called television drama "summer stock in an iron lung." Film people who feared the rapidly growing infant contemptuously named it "grade-Z movies with the fluffs left in." A few radio producers who foresaw the end of radio as an art whistled past the graveyard by referring to television as "bad audio with pictures."[1]

GOING LIVE?

An approach to television must start with the intrinsic nature of the medium. One of the characteristics for the specificity of television is its function as a relay medium, without separate aesthetic principles. Originally, all television broadcasts were live and television's essence was understood to be its ability to transmit events in "real time," rather than in a filmic capacity of recording events for later viewings. In this way, if television has, since the 1950s, long ceased to have direct-to-air transmissions so that even sports matches are prone to electronic manipulation, and the medium cannot be argued to be live in an ontological sense, it has remained ideologically "live" for a far longer period. Jane Feuer, writing in 1983, argues that the definition of television as an ontologically live medium continues to be the dominant method of thinking about the medium.[2] Rick Altman has made a similar observation: "Whether the events transmitted by television are live or not, the television experience itself is thus sensed as live by the home viewing audience. Just as the camera has to be on the spot to record a live news event, the potential spectator must make sure that her eyes are on the television when something important happens—or risk missing it forever."[3] A live production is not recorded and will not leave a durable presence as with recorded film. John Ellis points out that when watching television "there is hardly any chance of catching a particular program tomorrow or next week sometime."[4] Until comparatively recently, the content of television was ephemeral, a transient moment never to return. Television like radio before it shared the same basic assumption that content was outside the realms of memory and repetition: it possessed no lasting value. Consequently, in regards to the videotape revolution from the late 1950s, television avoided the temptation to be like literature, which can possess a value that transcends a specific historical moment. As a consequence of this live period and its legacy, early science fiction on television became ephemeral, marginalized, and lacking cultural status.

Yet, despite this ephemeral nature the belief that *some* television is worthy of sustained scrutiny and critical attention echoes Jason Jacobs, who, since the 1990s, has argued that there is a category of programs on television that "begin to demand the criteria that have to account for their excellence."[5] The demand for aesthetic evaluation is however complex, as he explains:

> The subjective basis of casual judgement is undeniable, but our judgements are not necessarily restricted to our immediate subjective response. As soon as we begin to reflect on that response, to think about it, to articulate it in conversation (with ourselves and others) and in our writing, we necessarily begin to impose an objective structure on it.[6]

Moreover, the creation of a canon on television is difficult not simply due to matters of taste and subjectivity but also because television is a medium whose extensive cultural form constitutes its essential characteristic as well as poses an ontological challenge. Television's enormous output complicates the claim that it has been a medium whose content is ephemeral. Rather its output is watched and rewatched because of syndication deals and reruns. However, it is the need to fill a daily schedule for an audience that is unprecedented in its size that has led to the centrality of categorization in television, which resists programs that can be considered to be unique. Instead, the industrial scale of television seeks to devise a scheduling pattern that creates an order and form that is specific to the medium. In this way, the production and consumption of programs are regulated by repetition if modulated by acceptable difference.

It is worth recalling Raymond Williams's concept of television's "flow" at this point. Williams first wrote about his model of flow in 1974. While working as a critic for British magazine *The Listener* (1968–1972), he had noticed that everything on television had become rather like everything else.

For Williams, the significance of the difference between broadcasting, including radio, and older cultural forms was not simply in the distribution of programs to the home but "a sequence or set of alternative sequences of these and other similar events [plays, meetings and stories] which are then available in a single dimension and in a single operation."[7] The move from "programming" to "flow" had been harder to identify in Britain because the public service ethic had preserved programs as a fixed unity and "the older concept of programming—the temporal sequence within which mix and proportion and balance operate—is still active and still to some extent real."[8] Nevertheless, according to Williams, the move from programming to flow meant that programs as coherent separate texts on television struggled to maintain a fixed unity.

Yet, within the flow, the segment, of which so much of broadcasting content is composed, has been its salient feature since its inception. Its main aim has been to not only attract viewers for the first time, either when they first switch on, or change channels, but also to hold on to them. The segmentation of narrative, according to British television theorist John Ellis, is television's own creation. Ellis argues that from the 1950s, television has relied on a segment of several minutes as its basic unit.[9] One illustration of segmentation was the fact that television often took the form of the serial or series, which became increasingly popular from the 1950s, rather than the single play or the anthology of self-contained programs. According to Andrew Crisell, with segmentation, the existing audience is retained because the "horizontal desultoriness" of the viewer; her inclination to keep browsing among all the channels available to her, is overcome by offering "vertical desultoriness: a constant change of stimulus within a single channel."[10] This form of engagement of television with its audience has led Jane Feuer to make a further observation that since the 1980s the typical viewing experience for television has been characterized as a rapid flow from one genre to another.[11] Nevertheless, regardless of the historical period, the ability to switch from one channel to another, which marks television's heterogeneity, does not simply make it harder to categorize separate programs within a particular genre, but also complicates our understanding of television form.

It is this sheer *versatility* of television that offers some of the clues to understanding the adaptation of older media forms to television. We can observe that older forms from film and radio were transported to the new medium, but they were also actively adapted to it. Consequently, different forms of entertainment, including the science fiction genre, were transformed by the new medium and became what may be regarded as "television" with its own modes and styles.

Although Ellis identifies segmentation as something that belongs to television, the narrative structure of segmentation in moving pictures was first visible at the cinema and in the type of film known as the "chapter play." The chapter play was a movie serial and its segmented narrative preceded television's, which arguably was able to inherit its form after the demise of its former rival in the early 1950s. The continuity-with-difference within the earlier movie serial achieved two outcomes in some of the first television shows that were to follow: First, it satisfied the viewer who by watching previous episodes realized there was an accumulation of information, adding to the value of the story. Second, it allowed viewers who had missed a previous episode to become reacquainted with events in previous episodes and encouraged continued watching week after week. In this way, television science fiction shows adapted the segmentation of the earlier chapter play, but the desultoriness of television identified by Crisell has permitted a greater fluidity of generic boundaries that had remained relatively distinct at the cinema.

The second main adaptation to affect television, in its formative period of the 1940s and 1950s, was the technology of film. Film became the method of recording a live program on US television in the 1950s. However, the methods by which the technology advanced and was finally accepted as the dominant paradigm for television shows such as *Science Fiction Theatre* was a process complicated by how the new live medium came to be understood in terms of its engagement with its audience, as well as economic factors. The adaptation of film to television reveals how institutional, economic, and social changes were to affect the form and mode of television. These were to become increasingly mutable once the technological capacity of television had expanded to include film recording and television ceased to be live.

During its formative period and because it was live, the essential nature of television was argued to be intimacy. The closeness of the image to the performance of the actors, unadorned and without the tinsel of cinema spectacle, could avoid the erection of a barrier between the screen and its audience. Television, unlike the theatre, lacks a stage, orchestra pit, or footlights to separate the performer from the audience. Instead, television was to inherit its bedside manner from radio to create the illusion of the proximity between the performance and listener/viewer. At the same time, it should be remembered that some television programs were and continue to be "audience shows" and are enjoyed by large masses of people gathered in some sort of auditorium or studio. Nevertheless, such big television variety shows, broadcast from a theatre stage, continue to give the viewer at home a feeling of being in the front row, as well as among the other members of the audience sitting in the theatre. The suggestion of an intimate method of viewing is strengthened by the notion of television as an example of a *domestic* medium. Crucially, it addresses its audience as individuals sitting at home rather than as *the* masses. It was within such complex conditions of production and reception that, in the 1950s, the genre of science fiction was adapted for American television. The purpose of this chapter is therefore to examine how the inheritances from radio and cinema were remediated for television. More specifically, the chapter looks at how the development of television from a live to a recorded (filmed) medium created competing modes and styles in the science fiction genre. Nevertheless, such competing codes should be understood and accepted as concomitant differences in presentation.

THE CHAPTER PLAY

From 1929 to 1956 over two hundred thirty film serials, or "chapter plays," were produced by the major studios in Hollywood, such as Columbia and Universal, and minor studios such as Republic Pictures, which had been set up in 1935.[12] In the United States, the Saturday matinee serial, or chapter play, appealed to a mainly juvenile audience, but to argue that the serial is necessarily a narrative form that appeals exclusively to the young or the unsophisticated would be mistaken. Many of the serials were family entertainment, enjoyed by adults accompanied by their children. Moreover, celebrated literary output has demonstrated what is possible within the serial form, including the serials of the romances of the cloak-and-sword variety by Alexandre Dumas, or the slowly unfolding science fiction plots and characterization of Jules Verne. It should be recalled that eighty of Verne's "*voyages extraordinares*" were, over more than forty years, serialized monthly in the *Magazine of Education and Recreation*. Nevertheless, the chapter play was a marginalized form of entertainment, a sideline, albeit a highly profitable one, to feature production. In the history of science fiction, it is probably true to say that the chapter-play continues to be a footnote, a juvenile auxiliary to canonical science fiction. But despite their minor status, they nevertheless had a durable legacy, as we shall see.

One reason for the critical neglect of the film serial has been the form's brisk pace, focusing on physical action, which makes it comparable to the comic strip rather than the literary text. The critical preference for the word over the image raises qualitative issues to do with adaptation, a subject dealt with later in this book, but it is sufficient to say that science fiction is often regarded as an "idea-driven" genre. However, the film serial's close relationship with the image, which was borrowed from the comic book, means that some science-fictional elements—telepathy, for example—are difficult to replicate on film. Conversely, robots, death rays, and rocket ships have been adopted successfully due to the ability of film to offer the spectacular. Annette Kuhn suggests that "the most obvious difference between science fiction as literature and science fiction as film lies in the latter's mobilization of the visible, the spectacle."[13] The mobilization of the visible and the spectacle of a futuristic technology within science fiction became evinced in the earliest serials at the cinema. In fact, the tendency for inventions to be a key choice within the narrative predates the specifically science fiction serial. For example, *The Black Box* (1915) is about a Holmesian detective, who invents several gadgets, including the means to see who is ringing him on the telephone. Outside the science fiction genre, these early serials were already setting a precedent that would appear in later science fiction: using inventions to initiate a conventional action plot.

One of the earliest science fiction film serials, *The Vanishing Shadow* (1934), produced by Universal, employs an inventor-hero who creates a device that when worn leaves only the user's shadow visible. The *Vanishing Shadow* also includes many other science fiction gadgets such as a robot and the Destroying Ray. The following year, *The Lost City* (1935) was about a "scientific wizard" called Zolok, who rules over a lost city inside Magnetic Mountain. The hero is an electrical engineer who, with his companions, journeys through darkest Africa and is set against Zolok. *The Lost City* uses a series of high-tech gadgets, including brain-destroying machines and TV monitoring systems. Although the mobilization of the visible exists within so many science fiction chapter serials, the gadgetry is essentially pseudo-scientific. It is deployed for its spectacle, but possesses a literal meaning devoid of any deeper significance, or any exploration of the possible social effects of the technology. Each new invention exists to advance a plot rich in action and is never fully documented. This becomes more obvious in the serials of the 1940s, in which flesh-and-blood heroes, such as Flash Gordon, are replaced with super heroes like Captain Marvel and Superman: the latter, as the son of Krypton, now equipped with his own attenuated "super-science."

The juvenile super-science of the earlier cinema and television serials would be absent later in *Science Fiction Theatre*, which was produced for American television in the mid-1950s. Shows such as *SFT* strove to be science fiction for adults. Physical action had been the foremost element in the serials of the cinema, but television could also draw upon another inheritance: radio's "theatre of the mind." In some ways, the film serials of the 1930s and 1940s mark the opposition between the cinema, with its mobilization of the visible as spectacle, and radio as a medium of dialogue/talk, which could appeal to a more cerebrally inclined audience. However, television is a complex medium, and such dichotomous oppositions, although superficially attractive, avoid a fuller understanding of the complex process of adaptation, as we will also see.

The film serials may have deployed a degree of physical action that would be less visible on television, but television continued to share a basic narrative structure with the chapter-play. The internal relation of serials on film with the serials on television was the possibility of narration across episodes. With its continuing, if segmented, narrative the serial form provided recurring characters and generated future episodes from basic thematic material. It was television's segmented narrative that later, combined with commercialization became the key to its dominant mode of addressing its audience, even if some differences remain between it and the older film serial. On a meta level, the move from single texts to a sequence or series of different programs would, as we have seen, create a "flow" on television, as envisaged by Raymond Williams.[14] But unlike the film serial, television had a domestic audience sitting at home rather than at the movie theatre, one that might watch an evening's worth of programs, rather than a weekly serial at the cinema.

Each of the studios produced serials with its own preferred style of chapter introductions. Many of the introductions consisted of a reminder of the storyline for the members of the audience who had seen the previous episode and a concise summary of the plot for those who had missed an episode. Sometimes, however, the introductions were not only concise, but also repetitive. Columbia tended to produce serials very cheaply and, for example, used several paragraphs in its introductions that were identical in every episode.[15] Only the last paragraph would be specific to the previous episode. The signature Universal chapter opening was scrolling text, similar to some of the Columbia serials. Besides introducing the characters, the opening was accompanied by what is often referred to as the "opening crawl" or a rolling text summary of the adventure so far. The opening crawl, which required some clever optical equipment, began in 1939, and was usually addressed to audience members who had missed the last chapter. Told in twelve episodes, *Buck Rogers* (1939) deploys the device of an opening crawl. Much later, George Lucas would pay homage to the serials by opening each installment of the six-film *Star Wars* franchise with an opening crawl, the most famous of which began: "A long time ago in a galaxy far, far away . . . " Some introductions used by the chapter-plays included a narrator, while others did not, using the opening crawl to substitute for the voice of the narrator. When *Flash Gordon* appeared on US television in 1951, however, game show host Jack Narz was employed to narrate the opening section, since the television distributor assumed that children viewing the show would be unable or unwilling to read the text.

Perhaps the most famous and fondly remembered film serial of the 1930s is *Flash Gordon.* It had originally been adapted from Alex Raymond's comic strip of *Flash Gordon*, which first appeared in 1934. Raymond had in fact been the cartoonist, and the writer, Don Moore, had joined in August 1935. Originally syndicated to the Sunday comics by King Features Syndicates, the movie rights had been acquired by Universal from King Features, which produced in 1936 the first of a series of chapter-plays starring the eponymous spaceman hero.[16] Universal, which had already established a reputation for producing horror and fantasy films as features, employed Henry McRae, the head of Universal's serial department, to produce four or more serial films a year, with shooting schedules of six to eight weeks per production. Most were budgeted at $175,000 to $200,000 each. *Flash Gordon*, according to most accounts, however, cost $350,000, and thus became one of the most expensive serials ever made.[17] The inflated budget indicates that the serial was intended for a wider audience, including adults, as well as its core of juvenile viewers. Nevertheless, the science fiction genre on the screen was identified increasingly with a younger audience. Without the need for elaborate characterization and conversation, close-ups were replaced by wide framing and medium to long shots. As in most serials, the filming in *Flash Gordon* generally avoided close-ups and the possibilities of editing, preferring mainly fights and chases shot using a system of long to medium framing, more suitable for a juvenile audience. Close-ups would have also revealed that stunt men often replaced the actors, although Buster Crabbe, the actor who played Flash Gordon, was an ex-athlete who was known to perform his own stunts.

Flash Gordon, like most serials, was released in episodes consisting of two ten-minute reels of film, although the opening episodes of some of the serials would run to three reels in order to better immerse the viewer in the plot. The divisions between the weekly installments formed a series of entertaining "cliff-hangers," a term dating to the silent era, when serials would leave characters literally cliff-hanging at the climaxes of weekly episodes. By leaving their characters in peril, the episode could add suspense, and resist any resolution of the narrative, helping to guarantee the return of the viewer, intrigued to discover how the hero had escaped certain destruction.

EARLY TELEVISION AND THE TELEFILM

After Universal's rights to the serials had lapsed, the *Flash Gordon* serials were among the first to be shown on television. The response became so favorable that Matty Fox, an ex-producer at Universal, approached King Features for the rights to launch a new TV series, entitled *Flash Gordon*. However, the series was not altogether a success. The low-budget show appeared flawed next to the original filmed serials, which, so as not to compete with the new series, were retitled *Space Soldiers*, *Space Soldiers: Trip to Mars*, and *Space Soldiers Conquer the Universe*. The premise of the TV series was summarized at the beginning of every episode in stentorian voice-over narration:

> This [exterior building shot] is the Galaxy Bureau of Investigation—where the great scientific and economic minds of the planets work in democratic unity to preserve peace, prosperity and equality among all men. But powerful and evil forces throughout the galaxy are trying to gain control of those GBI secrets by which the democratic balance is maintained.[18]

The idea of a "space patrol" was popular in other juvenile science fiction shows on television at the time. Gary Colville and Patrick Lucanio have argued that because science and technology were dehumanizing agents "the stories and characters reinforced a communal ideal rather than advocating an ego-driven political or social ethic."[19] The development of role models on television—essentially still regarded as a live medium—also presented the possibility of direct access to live happenings. If the stories were set on other planets, they were places that were being relayed by the miracle of television; the new form of communication for the postwar era. A tension existed between the sense of wonder created by the stories and the black and white images coming out of the orthicon camera, which suggests a simple relay and the lack of control of the production of the image. However, if television is live, it is a medium whose time-based nature does suggest immediacy with events unfolding in front of the viewer. In this way, juvenile audiences came to believe in their heroes such as *Flash Gordon*, *Captain Video*, and *Commando Cody*, and the shows becoming travelogues of their journeys into space at a time when, with the Space Race still in the future, no such authentic imagery existed.

The juvenile appeal of the film serial, with its emphasis on escapism and adventure, was also transported to television in the earliest science fiction programs, for example, *Captain Video and His Video Rangers* (1949–1955), *Space Patrol* (1950–1955), and *Commando Cody: Sky Marshal of the Universe* (1955). The DuMont Network pioneered science fiction for television by initially broadcasting *Captain Video* live five times a week, for six years, for young space enthusiasts. The eponymous character was "an archetypal hero. . . . He was Odysseus, Hercules, Hector and Achilles, but the sword was supplanted by the ray gun."[20] The fetishization of inventions, which had been thematically adapted from the cinematic serials, reappeared in the form of Captain Video's "Opticon Scillometer," a gadget that scanned through walls. *Space Patrol*, with its invisible rays became another imitator of super-science. Devised by the ABC network, it appeared on Saturday mornings.

Although the serial narrative structure had been transported to television, however, a degree of adaptation of the visual style occurred. One clear difference between the film serials and science fiction on television in the 1950s was that the latter was seriously limited by technological and budgetary restraints, employing special effects that were, if anything, inferior to those of the movie serials of the 1930s.

One major problem not only for depicting special effects but also for creating a sophisticated visual composition, was the complexity of broadcasting a show that was being shown live. Until they could be shot on film or recorded on videotape, television shows were broadcast live and often produced in a small studio, analogous to producing within the theatre. The ability to record and, moreover, edit a program still lay in the future. The first videotape recorders would be developed by Charlie Ginzburg and Chuck Anderson in the 1950s at the American corporation, Ampex, which made professional audiotape machines for radio broadcasters and recording studios. Charlie Ginzburg was to join the company in 1951 with the purpose of developing the by-that-time familiar tape sound recorder and putting pictures on tape. Eventually, the first reel-to-reel videotape recorder, the Ampex VR-1000, was unveiled in Chicago in 1956 at the National Association of Broadcasters (NAB) to an audience of TV engineers and station managers. By 1961, videotape had replaced true live production. Until that moment arrived, however, recording had to be done on 35 mm film using a process known as telerecording (Kinescope in America), which was a remarkably elaborate method that used a specially adapted, large, high-grade flat television screen and a film camera aimed at the screen shooting at a frame rate synchronized with the television frame rate.

> The [film] camera has to photograph both the odd- and the even-line scan on one frame of the film. To do this without showing the lines a mirror drum arrangement is used, and the film is run steadily through the camera (which has no shutter). . . . The recording is not made by a camera on the set but from a picture of the scene produced on a monitoring set with a large cathode ray tube. An extra deflection coil is wound on the neck of the tube, which causes the electron beam to wobble up and down at a high frequency and so spread the light spot over the whole of the scanning-line space.[21]

The image, as can be imagined, was poor; the process cumbersome and expensive. However, live broadcast was not practical in a country as large as the United States. In the 1950s, on the West Coast nearly all programs that were broadcast had been kinescope recorded after a showing on the East Coast. The purpose of kinescope recording was to overcome the problems of broadcasting to a country with four separate time zones. After quick developing and drying of the film, the "hot" kinescope would be delivered to the West Coast. In this way, it became common to broadcast repeated programs. However, the picture quality of kinescope recording was noticeably inferior to live television. One solution became not to record a live production at all, but produce a filmed drama using some or all of the techniques of moviemaking.

The debates about the associated problems of making films for television had already recognized the problem of the new medium supporting the costs normally associated with major film production. Bernard B. Smith writing in *Harper's* magazine in 1948 had explained:

> People will look at and listen to television programs for the same reason that they now listen to the radio: the television set is placed where it will form a part of the living habits of the American people. They will accept a much

> poorer level of entertainment in their homes than they will demand if they have to leave the house or apartment to attend a public performance.[22]

This notion of the domesticity of television invokes the idea that it attracts the distracted glance as opposed to the intense gaze. The smallness of the image on the television screen and the lack of the fidelity, as well as the resolution, of the film image will render television to be a secondary medium, inferior to the cinema. Consequently, some theorists have argued that whereas the film image engages the viewer's rapt, unwavering *gaze*, television elicits the glance: the viewer will look intermittently at the screen because of the distractions of the home.[23]

Such patterns of position and the flow of programs raise the question of who is at home, as well as when a program appears on the schedule. But the belief that because it was a domestic medium television would cater to the individual or group in the living room also permitted the idea that television was not being made at all for the mass audience of the cinema.[24] Unlike theatrical film, the earliest programs on television, transmitted live, were not considered to be a substitute for film because they could not match the perfection of film's unlimited retakes and the commercial limitations within the television studio affecting lighting and settings.

> As Programming expanded some networks turned towards the West Coast for their directors, feeling that films would provide added experience in the control room. Unfortunately, a great many of this type of personnel failed to survive the ordeal. This was largely due to a basic failure of the movie trained people to adjust to a technique which allowed for no retakes if one scene did not come off.[25]

As Bernard Smith argues, within a domestic setting, it was believed that the audience would be willing to accept other losses of fidelity such as the background noises of the studio, clumsy camerawork, fluffed speech, poor lighting, and inadequate sound quality.[26] If these problems could be avoided using longer camera rehearsals, there were however also limits on rehearsing facilities and the time that could be provided.

At the same time, the filming of television programs appeared to be prohibitively costly. As early as 1940, one estimate argued that because a motion picture cost from $1,000 to $35,000 a minute—that meant television could never compete with the cinema.[27] The problem was television had to fill a *daily* schedule, which could be several hours long. Instead of admissions to see a film on a pay-per-view basis, television was to become a system of continuous, "free" viewing. Scheduling, important in the commercial development of television, was to slowly appear in the late 1940s, and by autumn 1948 some of the first drama anthologies were to be seen at fixed times on CBS. Unfortunately, in the early 1950s, the new medium lacked large audiences and therefore the advertising revenue that might be expected if there had been sufficient audience size to raise the low prices of programs.

For the moment the major studios in Hollywood had been dissuaded from producing for television, although this was not a situation that would last long. Instead, the opportunity to produce programs shot on film was to be exploited by independent producers. Such producers as Fred W. Ziv began to enter an entirely new market that began to develop for television in the late 1940s and early 1950s: the telefilms. Parenthetically, a distinction has to be drawn between the telefilm and the telerecording or recording of a live television program on film. In the 1950s to 1960s, the telefilm was a term that was used to describe a television series that had been shot as a film and not recorded as a transcription of a live performance in the electronic studio. The telefilm used the standard techniques of film production, although it had to make concessions to the limitations of the television screen and the shorter production time allocated to a program compared to that of a feature film. Constraints on the telefilm would lead directors and producers to adapt visual and aural techniques from feature films to present a maximum of narrative information in as short an amount of time as possible. For example, they might use rack-focus to direct attention or segment conversations. The telefilm also tended to embody action and spectacle, as genre subordinated character to situation or plot. Finally, it could also be exported as well as shown in the United States. It was the invention of the telefilm as a form and mode specific to television that gathered pace in the 1950s due to syndication techniques pioneered by operators such as Fred Ziv.

Ziv was an advertising executive who had become the largest syndicator of radio programs by the 1940s. Unlike network programming, syndication allowed increased independent control of programming by local broadcasters, eschewing programming by network television controlled by ABC, CBS, and NBC. Soon the industry was buzzing with talk of residual rights, and these rights could be substantial, netting a producer more than the initial showing. Much later, Ziv explained:

> You know what the network's attitude is today. It was the same then. The networks knew everything, and assigned someone as a supervisor. They dominated the production of any show we put on the network, and that is one of the reasons why I, personally, preferred to put shows in syndication.[28]

Ziv's first program had been a soap opera, but by 1940 the Ziv Company-Radio Productions was expanding into other genres of radio programming. The equivalent of the film serials also existed on radio: programs such as the western *The Lone Ranger* (1949–1957) and the mystery series *The Adventures of Ellery Queen* (1950–1952). The purchase of the radio and television rights to the western *The Cisco Kid* (1950–1956), and its success with audiences led to the Ziv Company being lauded as the "largest of the open-end transcription producers in the country."[29] In 1948, with the formation of Ziv Television Programs under its new president, John Sinn, *The Cisco Kid* was on radio, television, and film. Sinn believed that film should become the method by which television programs would be syndicated to regional markets. However, the low prices for programs meant that production conditions for early telefilms could be challenging. *Business Week* would describe the speed of production at Louis B. Snader's Telescriptions studios: a sixty-minute episode could be produced on as little as a one-and-a-half day to three-day schedule.[30] For a successful producer of telefilms, Ziv admitted, "In the early days of television, we had to produce these things cheap. There's no question about it, and cheap is the word. Not inexpensive, but cheap."[31] Such constraints meant that the accepted Hollywood style of filmmaking could never be copied in its entirety, but had to be adapted to the new exigencies imposed by the pressures of producing material for the cathode ray tube.

Low budgets, tight rehearsal times, small casts, cheap sets, and the strict enforcement of production times all created a direct parallel with television's live transmission affected by the same restrictions. Without access to larger budgets, the telefilms relied more on rigidly formulaic modes of production and standard products than theatrical films. In other ways, too, budget constraints compelled low-budget filmmaking to become less cinematic than even the film serials, and more like the live television shows of the 1950s. A live performance demanded a high degree of professionalism, and a quick intelligence, for example, to make rapid decisions if something went wrong, such as a collapsing prop or actors forgetting their lines. Due to a show being live, the most important thing was to keep the program *flowing*. Actors had their marks and knew where to stand or else move to, and camera operators were ready with their angles. Live production depended on it being done continuously and without a break. But rather than the experience of live television being dichotomous to working on film, it prepared the actors and technical crew for the production of short-scheduled, low-budget telefilms.

Early telefilms would serve as a transcription medium in the same way that by recording on disc and, increasingly from the late 1940s, on audiotape, radio shows had been distributed to the hundreds of regional and local stations in the United States. The idea that new technologies possess medium specificities and can exercise considerable disruptive power on an existing commercial system, as already existed at the cinema and within radio, has to be balanced by how the new technology (television) formed a parasitical relationship with the earlier one (theatrical film), as well as others in print and radio. Rather than rigid boundaries between media, or their essential natures, there is often a fertile exchange between earlier and older forms and modes. Nevertheless, the adaptation of film, initially as a transcription medium, as well as sound for television enables a closer study of how visual and aural techniques became adapted and whether these techniques can be distinguished from earlier media.

The production of the filmed, half-hour television program did not escape criticism. In 1950, Ziv signed a five-year $100,000 lease with California Studios in Hollywood to produce television programming.[32] Ziv's operations were expanding, but opposition to Hollywood and films came from Jack Gould, a critic for the *New York Times* and one of the most influential daily critics of television. In "A Plea for Live Video," Gould would rail against the "dogeared films that Hollywood is turning out for television, the pedestrian little half-hour quickies that are cluttering up the facilities of even the best of networks."[33] Praise for live broadcasting continued when NBC president Robert Sarnoff, after quoting Gould, argued during testimony to Congress:

> Today, television broadcasting is at a crossroads: one fork has color signposts and points to programming created for the medium itself, with emphasis on live service. The other fork follows a detour to a reservoir of motion picture film, built over the last twenty years. At NBC we have carefully weighed the alternatives for the network and our owned stations. We have decided that television's future rests along the route we now chart. We shall continue our emphasis on live television, our fresh new programs designed for the medium, and on the development of color. We believe this is the way to maintain television's momentum and vitality.[34]

The fear was that network television programming had been reduced to "the lowest Hollywood denominator . . . a continuing flow of stale and stereotyped film product."[35]

Instead of the malign influence of theatrical film as it had been practiced in Hollywood, men like Sarnoff and Gould had continued to plead the case for the continued support of the "television play." As we have seen, the economics of television was leading from live production to shooting on film, and limits on production values. However, cultural productions on early television had been often initially modeled on live theatre, especially dramatic presentations. Television's "Golden Age" of drama consisted of the "anthology series," such as *Philco Television Playhouse*, *Kraft Television Theater*, *Goodyear Television Playhouse*, and *Playhouse 90*, among others. The anthology series consisted of different scripts by separate authors, which were presented every week, and expected to receive the greatest serious critical attention.

> The most successful "live" television play . . . would seem to be those which do not have much plot. In the half hour structure (make it 24 minutes at the most . . .), usually in two acts, a *single* basic situation in which the proper changes are played to a logical solution has so far been the best. If you have more than one situation, if you attempt a sub-plot, there isn't time to develop

> character; and if you don't develop character, the people aren't real and the viewer won't believe what he's seeing; and if he doesn't believe what he sees, he's wasting his time viewing. . . . "Believability," then is another fundamental of the best "live" television playwriting.[36]

Each week the networks would broadcast live original plays. Many little-known actors in New York, where the broadcasts would come from, began to attract national attention by appearing in these dramas—future stars such as Rod Steiger, Joanne Woodward, and Paul Newman. In December 1954, four of the top ten shows in the ratings were anthology dramas.[37] These dramas were defined by their naturalism and the realism of their dialogue, which could often be incisive, well-written, adult drama containing "serious" themes.

Yet, despite the importance of live program origination and a close study of the performer, it was not an essential element of television. Television programming had mainly inherited its program types from radio broadcasting, including the anthology mystery and crime series:

> Scripts should be prepared in such a way so that the viewer can follow the plot by listening, and not be required to remain glued to the set at all times to follow the action. On the other hand, the dialog should not explain every happening. In short, television-film writing should be a careful blending of radio and motion picture scripting.[38]

The growth of television form, specifically the telefilm, precluded television from fully adopting a paradigm from radio or theatrical film. Instead, television deployed a heterogeneous aesthetic to deal with contents that ranged across many different types of material, from live performance to filmed stock footage (inserted into the live performance, usually during a scene change), and even diverse forms such as animated comedy or a didactic report. Nevertheless, the dominant rhetoric that came to be created in the telefilm was cinematic and consisted of framing, distance, lighting, and editing. But on television, such rhetoric was affected by the substantially smaller screen size and became a modification of the modes of film technique.

The change to recorded drama, especially using film, which Jack Gould had identified with the exploitative commercialism of Hollywood, meant that, unlike a live performance, a drama could be repeated. In 1940, the expectation was that the reuse of film material would have limited appeal and commercial value. According to one commentator, "Rarely does a moviegoer see a film more than once. There is no reason to believe that the looker will consent to see a telecine transmission more frequently. Afterwards, the film must be relegated to the vaults."[39] However, television programs, shot as the soon-to-become-ubiquitous telefilm, either on 16 mm or 35 mm but mostly on 35 mm and occasionally in color were, due to the power of syndication, becoming valuable commodities to be bought and sold within a nascent but fast-growing market.

DISNEY AND EDUCATION

The move to telefilms programming by the networks was a response to its growing profitability by the mid-1950s. One of the most significant entries into the new market for TV shows was Walt Disney. The Disney studio had recognized television's possibilities of becoming a popular, mass medium, and had opened negotiations with ABC to form a funding agreement for its original theme park, Disneyland. The studio agreed to produce the TV show, *Disneyland* (1954–1958), for ABC largely in exchange for the network's financial backing of the proposed park. Each program for ABC would be linked to one of the themed "lands" that were to form Disneyland: Tomorrowland, Adventureland, Frontierland, and Fantasyland. *Disneyland* was to become ABC's first show in television's top ten. One of the first outcomes of the agreement would be the Tomorrowland-themed "Man in Space" series of episodes for ABC: "Man in Space" (March 9, 1955), "Man and the Moon" (December 28, 1955), and "Mars and Beyond" (December 4, 1957). In the shows, collaboration with Disney offered abundant cinematic possibilities in television, both visually and narratively. Of equal significance was the fact that the Tomorrowland television shows attempted to develop a rhetoric that would be rooted in science fact, eschewing the super-science and gadgetry of the film serials and most of the earliest science fiction TV shows. Scientific accuracy and detail would become paramount in the first of the Tomorrowland trilogy produced by Disney.

The use of technical excellence and spectacle would differentiate "Man in Space" from earlier shows such as *Captain Video*. For the first time, science fiction on television would be visually captivating not simply due to its escapist thrill, but because it was plausible. The first in the trilogy, "Man in Space," is a mix of animation and documentary techniques, whose wider aim is to demonstrate that television is more than a source of entertainment. Scientific and technological issues about the building of a rocket to take men to the moon are presented in an accessible and popular way. It opens by using voice-over to introduce the viewing audience to Walt Disney. The avuncular Disney sits on his desk rather than behind it, and casually plays with models of rockets, which appear later in the show—including and particularly recognizable to the avid enthusiasts of space travel of the time, the familiar German V-2 rocket. What has been considered to be fantasy will soon be reality because of the advancement of science, explains Disney. "Man in Space" has altogether five narrators: the first two are Disney and his long-term employee and chief animator, Ward Kimball. The last three are German scientists.

The function of Kimball is to build a program for the viewer out of filmed subject matter. This is the most "filmic" section of the program. The editing of different stock footage offers a discontinuous use of space that, for example, begins with a panning shot of Kimball's office before embarking on an animated history of the use of rockets and early space travel ideas. The account of the history of rockets begins in medieval China and forms a montage of explosive effects in the shape of a dragon and stylized shapes and patterns before shifting to a discussion of one of Newton's laws of motion using an animated dog. Next, it is followed by a recap of one of the first important books about space travel, Jules Verne's *From the Earth to the Moon* (1865). Interestingly, the show moves onto Georges Méliès, the pioneer filmmaker, and one of the first films about space travel, *Le voyage dans la lune*, made in 1902. Clips from other landmark science fiction films are also included such as *Frau im Mond*, directed by Fritz Lang in 1929, as well as actuality footage, which is accompanied by a music track that can be described as "vaudeville."

The hybridity of the footage suggests the compelling nature of the first section of "Man in Space," and Kimball's desire to popularize science. However, as the footage progresses, there is an awareness as to whether the audience will find the later "live" footage only an uninteresting and impoverished version of the filmic. Over footage of the launch of the V-2 rocket, Kimball narrates how within Nazi Germany there was an accelerated rocket program. Even from today's perspective when space flight has become normalized by its familiarity within both fact and fiction, the sight of the V-2, the forerunner to modern ballistic missiles, appears impressive. The excitement generated by the rocket is acknowledged by a montage of V-2 rockets being launched. The editing serves to dramatize the rocket and ends with a large explosion as one of the rockets detonates on the launch platform. The section concludes with a further extended montage of the sight of captured V-2 rockets, as well as other American rockets, being launched from the US Army's testing site at White Sands, New Mexico.

"Man in Space" is mediated to the audience by the techniques and persona of the presenter Kimball, who offers an advanced visual animated art. The burgeoning economy of repetition, as programs were shown more than once, and the shift from live television to the telefilm would mean that "Man in Space" represented the embrace of the cinematic paradigm for television. Kimball's cinematic discourse is largely created to impart information about the development of the rocket, but exists mostly as entertainment. The popularization of science for a television audience required the audience to continue watching, thus Disney wanted to avoid having the show become too didactic. The narration by Kimball is initially lighthearted before becoming more overtly dramatized. Disney's suggestion had been that the programs combine both

> comedy interest and factual interest. Both of them are vital to keep the show from becoming dry. You need a good balance to keep it from becoming too dry or corny. We don't want to compete with Sid Caesar or do that type of thing. We want to do something new on our show.
>
> We are trying to show man's dreams of the future and what he has learned from the past. The history might be a good way to work in a lot of your laughs. People laugh at inventions of the past . . . because with the inventions and the real progress of science today, people feel superior. . . . [But] we have to watch it so the material doesn't get corny. . . . I think this parallels the "True-Life Adventures"—[presenting] facts, and opening up this world to people.[40]

The use of Kimball, a nonscientist, as a narrator demonstrates that the aims of education and entertainment are intended to be complementary within the show. However, the animation and stock footage of the rockets is almost too compelling to successfully avoid the idea that the filmic and the "live" are not in opposition to one another. The appeal of entertainment followed by a segment that is didactic, using hard science to explain rocket propulsion, demonstrates the difficulty that Disney was faced with by combining the two.

While being lighthearted, "Man in Space," at different moments, offers a serious mode, which is distinct from the animated comedy. Unlike cinema, television could edit in images and sequences to alternate live sequences with filmed inserts. The mode of address and the type of discourse in "Man in Space" was to be heterogeneous, and these distinctions mark content that is recognized as "real" and content that is fictional. "Man in Space" demonstrates how the visually dynamic, the pictures full of incident, are less serious than the more "live" sequences.

"Man in Space" was also planned to look like science fact rather than fiction. The trend to science fact had begun earlier with adult shows such as *Tales of Tomorrow* (ABC, 1951–1953), and the CBS science fiction anthology, *Out There* (1951–1952). *Tales of Tomorrow* had been produced in cooperation with the Science Fiction League of America and had involved the TV rights to adapt some two thousand stories by science fiction authors. It had been decided that the key to success was to keep the action contemporary but extraordinary. The show was to avoid bug-eyed monsters (BEMs), futuristic sets, and costumes. Rather than a telefilm, the stories were to appear more as a live stage play. Nevertheless, *Out There* was one of the first series to combine live action with filmed special effects. However, "Man in Space" was to combine not simply "live" action with special effects but three types of moving image: "live action," animation, and inserts of actuality footage of the testing of rocket technology by the American military. In fact, the interplay of these different modes in one program was to work against the myth of actuality on television. Instead, unlike earlier television, characterized by its ephemeral nature, "Man in Space" becomes a commodity that can be repeated and treasured due to its high cultural worth. At the same time, the footage of the direct address by Disney, Kimball, and finally the German scientists creates the illusion of instantaneousness, if not quite the disposability about a topic limited to the moment when it was first shot.

The use of direct address is indicative of the dual role of the program, hovering between education and entertainment. Ward Kimball and the animation had provided the entertainment, but the German scientists were to be educational and prove that television was capable of being a medium of serious exposition. After the filmic montage, Willy Ley, the first scientist, is introduced by Kimball, who is described as a "rocket-historian." Ley is the eldest of the three scientists and takes precedence. Fulfilling his pedagogical role, he stands in a space that has been dressed as a classroom. His delivery varies between direct address[41] and answering questions from the young American animators-technicians in his class. The exchanges are scripted and clearly not spontaneous, but, in many ways, because the classroom is situated in the nonspace of the studio, and within the anonymity of this type of space, so different from anything "real," the stiff diction is rendered more acceptable. Ley sounds authoritative and omniscient, and it is in his role as an educator that the confined studio space allows for direct questioning from the technicians which draws upon Ley's scientific expertise. At the same time, his omniscience is not neutralized by the use of an anonymous studio, because he participates in the film as a character; he is one of the German *Experten* who will make possible our adventure into space. His status remains authoritative but his German accent, the accent of a recently defeated, but technologically advanced enemy, is humanized by his appearance. Dubbed over an additional animated sequence following his live-action appearance, his voice becomes disembodied as it narrates the testing of space travel using a new satellite, before the sequence completes and the film cuts to a shot of the movie projector on which the film we have been watching is playing. Here exists an attempt to demystify the film apparatus and remind the audience that although the animation and stock footage can help us imagine what the conquest of space may be like, it is only speculative. However, the live medium of television continues to speak more directly and, as a consequence, the appearance of the scientists is rendered more believable.

The German scientists as narrators fulfill the role, not only as educators, but also as commentators, who are the "on-the-spot," and able to inform us about the possibilities of the exploration of outer space. In their roles of educators and pundits, Ley, Heinz Haber, and Wernher von Braun make use of diagrams as a descriptive form of communication to elucidate the commentary about the impending space race that would dominate the next decade. Both roles are collapsed by their status as narrator, which also highlights the importance of the television presenter to the degree that this section of the program becomes an extension of the personality and style of the German scientist.

The structuring of the images of Dr. Wernher von Braun, a handsome, charismatic aristocrat who had been instrumental in the Nazi program for the V-2 rocket, is an adaption of live television's ability to transmit events as they happened: a sense of objective reality, unmediated by the camera. When von Braun first appears in "Man in Space," he is casually sketching (rockets?) with two of his younger subordinates. He is also situated less in the nonspace of the classroom-studio which we saw earlier, and more in his actual workplace, before there is a cut and he addresses the audience. Live television communicates in images to reproduce the forms of a live actuality and the illusion of formal transparency. However, the group of German rocket and space experts such as von Braun uses a form that is not quite live actuality, but rather a media report as he discusses the mechanics of exploring space. The effect is of the program taking the form that the audience values television for: its ability to present the direct experience of the scientists in their office, while adding commentary. Television's power unlike feature film is to "capture reality" and transmit it into the living room. The fact of the space race is unfolding before our eyes.

SCIENCE FICTION THEATRE

In the 1980s, John Ellis argued that British television programming lacked a narrative image.[42] For him, broadcasting had not developed the institution of the narrative image because television had not created the experience of the

Figure 1.1. Wernher von Braun in the role of educator and pundit.

single text, but was the *flow* of an evening's worth of programs, wherein the distinctive characteristics and a sense of differentiation between items on the schedule was missing. The narrative image in the cinema was in the form of an anticipatory reply to the question: "What is this film like?" But it was missing from television. Earlier science fiction anthologies such as *Tales of Tomorrow* had been confronted by the obstacle of the lack of what might be called a "narrative image." *Tales of Tomorrow* adapted stories from Jules Verne, H. G. Wells, and Oscar Wilde and also included original stories. Its range was characteristic of a science fiction anthology rather than a serial or series, which might use recurring characters and often continuous storylines. Consequently, there may have been less standardization and formularization of stories and character type, but the key anticipatory question of what this show is like, if we take into account Ellis's argument, remained unanswered. Yet, the anthology series, *Science Fiction Theatre* (1955–1957), produced by the Ziv Company, would remedy this by the contribution of a narrator at the beginning of every episode.

Science Fiction Theatre was produced by Ivan Tors, a Hungarian who had been a premedical student, before leaving for America and eventually arriving in Hollywood. Tors's interest in wildlife, science, and education was reflected in the films that he produced. He was an avid reader of *Scientific American* and would get ideas from it to turn into stories. However, Herbert Strock, a director on some of his films, believed that "Ivan had a fetish for doing scientific things that an audience wouldn't understand. He would lose the human element . . . and get into technicalities."[43] Tors's interest in space exploration had led to films such as *The Magnetic Monster* (1953), *Riders to the Stars* (1954), and *Gog* (1954). Props that appeared in *Riders to the Stars* would be seen again in *Science Fiction Theatre*, including microphones, speakers, fake computers, and the charts hanging on the wall. In July 1954, Tors began negotiating with Ziv and promised that the proposed science fiction anthology would avoid the type of stories seen on shows such as *Captain Video*. Instead, there would be a strict adherence to scientific fact and dramatization based on scientific enquiry. To this end, several press releases and magazine articles claimed that Tors had been granted a budget of $1.5 million to shoot on location at Air Force bases, universities, and private laboratories. However, if the series could not be shot on stage, or on location within an hour's drive from the studio, stock footage would be used.[44]

The key innovation within *Science Fiction Theatre* was its narrator, Truman Bradley. Bradley possessed the physiognomy of a well-scrubbed executive, and had been a radio announcer, working on *Ford Sunday Evening Hour*, an hour-long music series, on the CBS affiliate, WBBM. A culturally well-regarded series, it confirmed Bradley's skills as a slick radio announcer. Bradley had also been capable of reading the news for *American Family*

Soap and *American Family Flakes*, but soon revealed that he was not limited to news, but also skillful at commentary, and he was employed to narrate seventy-eight episodes of *Science Fiction Theatre* from 1955 to 1957. The pilot, "Beyond," was filmed in July 1954, and its opening and closing segments, featuring Bradley, were filmed separately from the dramatic stories. Bradley's on-screen narration was not filmed until September 11, which included his off-screen narration for the pilot and the second episode, "YORD."

Each episode of *SFT* began in a laboratory inhabited by Bradley. The laboratory was dissimilar to the non-space of the classroom-studio observed in "Man in Space," and more like the realistic space of a working laboratory. It also eschewed any inferences to the super-science and the types of electrical equipment visible in the older Universal serials. The setting would not be that of the mad scientist but of the technician-executive, who possessed the cool professionalism of the rocket engineers in "Man in Space." The accompanying music at the beginning of each episode was a brass fanfare, and it played as the camera ranged over the laboratory before the title credits appeared. Unlike virtually all Hollywood films until the 1960s, which consisted of initial titles against a fairly static background, television openings and titles had to immediately "hook" the viewer. The pieces of laboratory equipment therefore had to be sufficiently intriguing to dissuade the viewer from switching to another channel. In the first episode, the opening sequence stops on an empty chair, further intriguing the viewer about the source of the narration, before cutting suddenly from the empty chair to a close-up of Bradley. Although the cut serves a dramatic purpose, a more prosaic reason for its inclusion was that the film camera would not have been equipped with a varifocal lens. The zoom lens had been in use since 1947 and had enabled a live studio camera to range fluidly from close-ups to long shots in one take. However, within the film industry, cinematographers continued to rely on prime lenses, due in part to their criticism of the possible distortion of space using the zoom lens, which was inconsistent with accepted Hollywood technique.

The persona of Truman Bradley was important in helping to establish the show's appeal. His appearance and demeanor reinforced the sense of an executive or technocrat, and the controlled voice that had been trained for radio continued to emphasize his slick professionalism, which demanded to be taken seriously. Although television was a medium that had struggled with the notion of its suitability for detailed exposition, Bradley's omniscience, humanized by his avuncular appearance as well as cultured persona, made for a sincere and affable form of presentation capable of being trusted. The role he played recurred throughout the series. Bradley would open each program with the suggestion that each story was fiction, but the distinction between fact and fiction, due to advances in science, was that the fiction was plausible if not yet possible. *Science Fiction Theatre* was a theatre less of possibilities than of what might be plausible in the not-so-distant future.

At the end of each episode, Bradley would promise to be back with "another story from the world of fiction and science." In this sense, Bradley's role can be compared to that of Don Herbert, the host of the popular 1950s children's educational scientific program *Mr. Wizard*, and Lynne Poole, host of the low-budget but critically acclaimed series for adults, *The Johns Hopkins Science Review*. Moreover, in keeping with his demeanor of a technician-executive, Bradley fulfilled the role not only of a diegetic narrator and presenter, but also a resident manager in front of a "live" camera. Although the entire sequence had been filmed, the direct address suggested an unmediated, straight form of transmission. The sequence would also be mostly shot as a single take, rather than by using edits, as if shot from a camera in a television studio, with only the use of a few cutaways.

This sense of watching the assembly of a broadcast process familiar from live television is strengthened by the use of explicatory footage to help the audience understand a particular point being made within the commentary. The message of the speech from Bradley in the first episode is to not trust our senses because they are limited as to what they can detect and can prove to be fallible. He offers proof of this supposition by placing a ball bearing between two giant magnets and demonstrating how the invisible force will suspend the ball in midair without any apparent visible means of support. Next, science has devised incredible apparatuses that can, for example, film a bullet, normally invisible to the human eye, passing through a television set using a high speed camera. When Bradley fires a gun at a television, it shatters and he tells the audience that they saw him hold the gun and witnessed the impact when he struck the target. But, he asks, did they actually see the flight of the bullet? He answers his own question—"Of course not"—and he goes on to explain that just because our senses do not perceive certain things it does mean they do not exist.

The fascinating possibility that we live within a reality delimited by our senses and incapable of perceiving everything that might exist surrounding us is dramatized by inserted footage of a moving bullet recorded using a high-speed camera. It shows how the speeding bullet creates air turbulence before colliding into the glass screen of the television, which explodes on impact. The sequence has two effects. The first is to underline the scientific basis of the program by showing us a filmed sequence that offers images that demonstrate rather than simply dramatize. In this way, the sequence confirms the purpose of *Science Fiction Theatre*. The second is, however, the added revelation of the drama that is intrinsic in science. By using a skillful blend of didactic material and dramatization, the implication is that death rays and the other paraphernalia of the "super-science" of film serials are also less intrinsically interesting than anything true science can offer. The inventions within the film serials and the later juvenile science fiction programs on television are consequently less dramatic than the *ideas* being offered by *Science Fiction Theatre.*

Shot in color, although broadcast initially in black and white, *Science Fiction Theatre* offers what can be described as the craft of the film industry, unlike the small blue-gray images of the early television sets of only a few years earlier. The special effects technician was Harry Redmond Jr., whose father had been special effects supervisor for RKO Studios. They had worked together for *The Secret Life of Walter Mitty* (1947), before Redmond Jr. had been in charge of effects for the science fiction film *Donovan's Brain* (1953). It was Redmond's balanced magnetic field holding a ball bearing in midair that Bradley demonstrated in "Beyond." These effects to create interest in and fascination with science and the images used to dramatize them tended to be less complex and detailed than those produced for high-budget feature films with more expensive production values. But the weekly repetition of a television series would mean that it would continue to have a relationship with the audience in ways that further answered Ellis's anticipatory question, "What is it like?" The laboratory equipment and new devices each week would serve to create a narrative image that would overcome production limitations and the standardization of the overall format of each episode, including using stock characters—the doctor, the scientist, the policeman, the military man, and so on.

The pilot episode, "Beyond," was conceived by Ivan Tors. It concerns Major Fred Gunderman, a US Air Force pilot who ejects from his experimental jet when it appears he will collide with an alien spaceship. Once he arrives back at his base, the authorities are skeptical of his claims, and refuse to believe him until an important piece of wreckage seems to indicate that his story is true. As the story begins, Bradley, who has already established an affinity with the audience as the narrator in the laboratory, is also able to create an affinity between the audience and characters by introducing the characters. These include Dr. James Everett, the scientist in charge of research at the base, and General Troy, who represents the Defense Department, as well as Fred Gunderman himself. The distinction between the narrated segment and the dramatized story is made clearer by the decision not to overlap Bradley's monologue with the story itself. In this way the series owes something to wartime training films, with an omniscience suggesting a semidocumentary approach. The approach, which had been familiar in training films and US government documentaries from the mid-to-late 1940s, meant that episodes of *SFT* were frequently based on case histories or actual true stories, and they relied on location shooting in order to lend them credibility and authority. Nevertheless, despite its insistence on authenticity, *Beyond* uses conventional stock characters and a typical military hero in Gunderman. The story is told using a standard narrative structure, including exposition and action, before concluding with the resolution to offer a satisfying neat ending that Gunderman did, in fact, see a "Flying Saucer."

Variety, in April 1955, reported that it felt that the semidocumentary approach in *SFT* had been mishandled. Bradley's lectures, it concluded, left too little time for the dramatic stories and insufficient buildup of characterization and situations in only twenty minutes. Partly as a consequence of this need to compress events on television within the half-hour format, *Variety* argued that the omission of the key sequence when the pilot sees the saucer eliminated suspense and tension.

> The Fred Ziv–John Sinn film factory has taken the semi-documentary approach with this series . . . and to judge from the first program in the series, "Beyond," the approach is almost an impersonable [*sic*] one, relying more on the content than anything in the way of ordinary dramatics. . . . Documentary style used serves to obliterate anything like the development of character or situation. . . . Direction is standard, but a ring of reality is given the film through some on-location shooting at an Air Force base.[45]

In many ways, the tension between ideas and drama on television had also affected the trilogy of space exploration films from the Disney studio. J. P. Telotte identifies this shortcoming as being due to the transformation of the real into the cinematic. Archival footage is re-created to produce a comic hybrid that would include a set of comic types that were the staple of that era's domestic comedies.[46] However, the style in *Science Fiction Theatre* is restrained, never deploying the same visual excess, which in this instance can be described as cinematic as opposed to the heterogeneity of the televisual form.

In episode 37, "Are We Invaded?" the series returns to the idea that flying saucers are visiting Earth. Seth Turner, a brash young journalist, and his fiancée, Barbara Arnold, witness a UFO that takes the form of a fireball in the sky. He argues bitterly with Barbara's father, Dr. Arnold, a distinguished astronomer, who dogmatically rejects the existence of flying saucers. Seth is trying to establish himself as a TV news reporter by producing documented evidence in favor of the existence of flying saucers. A mysterious Mr. Galleon, who also witnessed the fireball, wishes to corroborate Seth's findings, but fails to appear. After Dr. Arnold's demonstrations have proved conclusively the impossibility of a flying saucer, an envelope arrives from Mr. Galleon, which contains an image of the solar system that could only have been taken from space.

The episode is interesting because, unlike "Beyond," it thematically rejects the power of the demonstrative image, which television has used as a site of fascination. Here Dr. Arnold remonstrates that Seth has been a victim of an optical illusion. However, Seth will seek out people who have seen flying saucers in ways that are similar to television seeking images and people whose eyewitness testimony can be used as evidence. In fact, Seth aspires not to be a reporter writing for the newspapers, but rather a television journalist , obtaining the highest level of professional status. The implication is that television can establish the "truth" by using the literal power of the image in a way that newspapers cannot do. The audience will observe and witness the images and make judgments about what they are watching—the people, the places or events unfolding as they happen. In this way, there is a direct correlation between the image on television and the "truth" of the events.

Seth films and interviews several people about what they have seen, including, somewhat quaintly, a minister as well as an airline pilot, a qualified aerial observer, an army sergeant who is a former chief of radar operations, an airways inspector in charge of control tower operations, and finally a retired air force captain. The last of these takes a polygraph test administered by an internationally known operator, who explains that in his expert opinion the captain did see unidentified flying objects. The cumulative result of these interviews is to underline the strength of witness testimony, which television has made, in news programs, one of its chief defining characteristics. Much of this strength relies on the "liveness" of television, which again can be argued to have been inherited from radio. On radio, the comprehension and interpretation of the object or event that is being described can allow for a greater independence of the audience, as they are invited to see in their own minds the thing that is being described. This creation of immediacy in television is discussed more fully in chapter 2, but here we note television's ability to produce live images or images that appear to be witnessed live also suggests the audience's engagement to be one of surveillance, or the gathering of evidence. Within "Are We Invaded?" there is the dramatization of this particular mode.

However, at the same time, it must be remembered that such images are not live but recorded in the form of newsreels shot on location. Seth records his interviews using a lightweight film camera, which is seen in the episode standing on its tripod, being used to record one of the interviewees sitting down in the studio. Nevertheless, this form of "direct" representation of reality demonstrates how the show, and television generally, is able to engage with its audience. The voice-over narrative, which explains who each of the interviewees is adds to the dramatization of the footage as it continues to describe the encounters each of the interviewees claims to have had with the flying saucer. The finished documentary is shown to Doctor Arnold. However, he claims that while the documentary is interesting it is also misleading. The power of witness that television provides is demonstrated to rely on the vulnerability of human testimony, which is flawed due to its ignorance of scientific facts. The passion that has driven Seth to attempt to prove flying saucers do exist is contrasted with the cold objectivity of Doctor Arnold, who relies on "fact" rather than "imaginative opinion."

CONCLUSION

As the medium of film was adapted to the electronic medium of television, certain stylistic changes were called for to make each show more suitable for a medium that in its formative years was still watched on a small screen. As we have seen, long shots were avoided, to be replaced with close-ups and medium shots. The sense of intimacy with the actors that this provided was an intimacy that also reveals a complex mode within television. The problems associated with a live performance were overcome by the ability to shoot retakes once recording on film was possible. Yet a sense of intimacy continued to be deployed in the form of voice-over and the re-creation of a live mode on film. Disney's "Man in Space" combines three modes—live action, animation, and actuality—but *Science Fiction Theatre* was to use only two. Truman Bradley was to provide the type of narration that might be found on radio as he began each story in the first person and linked his narration to the dramatization of a scientific principle that he demonstrated in the laboratory. This form of narration combines images that can demonstrate as well as dramatize. In these ways, the textual limits of the science fiction genre are continued as well as expanded, as it was adapted for television in the 1950s.

The development of the telefilm in the 1950s expanded the technological capacity of television to present images to the viewer at home. As film began to be used to make recordings of live programs, the televisual image was complicated as new methods were devised for presenting style and action in the science fiction series. In many ways, the use of film challenged the idea that television was a live medium. Instead, the science fiction genre developed a "narrative image," which replaced the live image with one possessing a greater visual distinction. The textual boundaries of the science fiction genre and its use of a distinct iconography became more cinematic in its ability to use images that were able to represent visual style and action.

The development of the genre as it was adapted for television was to demonstrate the dramatic character of science using several distinctive modes of viewing. The combination of didactic material and dramatization enabled a series of different programs that could satisfy the acute interest from the general public in the developing space race and other scientific possibilities and marvels. The presentation of stories with themes that might be scientifically possible in the near future connoted a new authenticity, as well as a seriousness that would be suitable for an older audience, rather than the juvenile audience that had been served by the older chapter serials. The

demonstration of technology in its various guises, but usually associated with the military as America fulfilled its role as the new postwar superpower, required Ziv Productions to fully cooperate with the military to show that its dramatization of America's technical prowess had been shot realistically. As Lucanio and Colville explain: "At mid-decade, television was moving away from being mere entertainment; it was beginning to take itself seriously as an advocate of public policy. . . . The television industry began to see itself as an emerging voice for social change."[47] But what sort of change would science fiction on television advocate on the other side of the Atlantic in Britain? How such a demand for social change would also become a demand for the fresh expansion of the storytelling abilities of television within science fiction is the subject of the next chapter.

NOTES

1. Stasheff et al., *The Television Program*, 3.
2. Feuer, "The Concept of Live Television."
3. Altman, "Television Sound," 45.
4. Ellis, *Visible Fictions*, 111.
5. Jacobs, "Issues of Judgement and Value," 431.
6. Ibid.
7. Williams, *Television*, 87. The equivalence of television with the older broadcasting form of radio is described by Williams, but lies outside the scope of the thesis.
8. Ibid., 89.
9. Ellis, *Visible Fictions*, 112.
10. Crisell, *Introductory History of British Broadcasting*, 282.
11. Feuer uses the series *Hill Street Blues* and *St. Elsewhere* as examples of genre hybridity. See Feuer, "Genre Study and Television."
12. Vahimagi, "Fantastic Cliffhangers."
13. Kuhn, *Alien Zone*, 6.
14. Williams, *Television*.
15. Backer, *Gripping Chapters*, 110.
16. Raymond's comic strips were often a source for the film serials, including *Jungle Jim* (1937).
17. Kohl, "Flash Back to the Future."
18. Quoted in Luciano and Colville, *American Science Fiction Television Series of the 1950s*, 116.
19. Ibid., 10.
20. Ibid., 8–9.
21. Morgan, *The True Book about Television and Radio*, 88.
22. Smith, "Television—There Ought to Be a Law," 37.
23. Ellis, *Visible Fictions*, 137.
24. There was a plan for theatrical television. In the 1940s, Paramount, Twentieth-Century Fox, and Warner Brothers were active in its development. See Gomery, "Theatre Television."
25. Hillman, "More Than Meets the Eye," 176.
26. Smith, "Television—There Ought to Be a Law."
27. Porterfield and Reynolds, *We Present Television*, 29.
28. Quoted in Broughton, *Producers on Production*, 18.
29. Grams, *Science Fiction Theatre*, 12.
30. Boddy, *Fifties Television*, 72.
31. Ibid.

32. Grams, *Science Fiction Theatre*, 14.
33. Boddy, "The Studios Move into Prime Time," 28.
34. Robert Sarnoff, quoted in Salent, Fisher, and Brookes, "Functions and Practices," 587.
35. Boddy, "The Studios Move into Prime Time," 29.
36. Roberts, "Writing for Television."
37. Wheen, *Television*, 102.
38. Quoted in Lafferty, "Television," 293–94.
39. Western, "Television Girds for Battle," 5.
40. Watts, *Magic Kingdom*, 310.
41. Not quite direct address, but a very close approximation of it for the audience watching the show.
42. Ellis, *Visible Fictions: Cinema, Television, Video*, 30.
43. Quoted in Zubatkin, "Strock Footage."
44. Grams, *Science Fiction Theatre*, 61.
45. *Variety TV Reviews*, April 13, 1955.
46. Telotte, *Mouse Machine*, 115.
47. Lucanio and Colville, *American Science Fiction Television Series*, 7.

Chapter Two

Screen Wars

BBC Television and Quatermass

> A steady progression is going on. More TV shows are being put on film; the difference in production costs is negligible, the general convenience is greater, the fear of fluffs and errors is eliminated, and the final product can be shown and re-shown without any loss of quality in all the different time zones of the United States. So more and more TV production is moving from New York to Hollywood.[1]

> Television . . . will be developed by the BBC as an instrument of education and information as well as public entertainment. . . . It aims to use television as an additional means of helping to bring about an informed democracy and generally to raise public standards. A compromise with the film industry would gravely damage this work.[2]

INTIMATE TELEVISION AND THEMATIC EXPANSION

In Britain, both before and after World War II, views on the purpose of the new medium of television were mixed. If sound broadcasting had begun as little more than the universal eavesdropper, then television had been the universal eye at the keyhole. The radio microphone, which had been placed at the boat race, horse race, or music concert, led to the assumption that television would continue to offer a relay device that could communicate events happening elsewhere. The presentation of the radio program had been essentially a variation on outside broadcasting and its main problem had been an engineering one: the satisfactory placement of microphones. Television, like radio before it, would risk being a medium that emphasized its technical nature at the expense of its aesthetic and creative potential.

Val Gielgud, the British Broadcasting Corporation (BBC) Director of Drama (1946–1952), wrote in the *BBC Quarterly* (1947), that television drama was not sure of its target: "Should its principal aim be that of *illustrating broadcasting*?"[3] It revealed the natural inclination to see television as concordant with radio. Cecil Madden, the BBC's first program planner and producer, writing in 1948, again reflected the uncertainty of the role of television, "Some people ask where television is going, whether it aims to be a photographed stage play, a competitor to the film, or an illustrated broadcast. The truth probably belongs somewhere between them all."[4] The notion that adaptation is additive and might incorporate new visual and aural methods of representing character and situation did not materialize as a demand for a "creative" television with its own distinctive methods of production. Instead, television was in danger of becoming relegated to live transmission and relying on older forms and modes borrowed from radio or the theater.

In many ways, it was a logical approach to adopt. The main appeal of a concert, for example, is to hear the music, rather than watch the members of the orchestra. Similarly, in an interview, sound is paramount rather than images of the interviewer or interviewee. On the other hand, drama often relies on the image for its meaning, and the spectator requires the visual cues of the play—the hiding of secret papers, the kiss on the heroine's cheek—to fully comprehend the action. Therefore, drama, perhaps more so than other forms of broadcasting, requires careful adaption. The possibility of adaptation suggests the deployment of the professional expertise of the producer, whose knowledge of the specialized technique of the early television studio was to become the construction of "immediacy."

The aesthetic of immediacy is central to an understanding of British television because it persisted long after the technological necessity for it had been removed. In its formative period up to about 1960, television was to create distinct forms of representation but the effect of these on audiences was to signify directness and spontaneity so that the authenticity of images on television became one of the characteristics of the specific forms of realism in television drama. Earlier, although some inexpert producers had tried to dramatize too many programs on the radio, other types of programs, for example, music or variety comedy, continued to use studio audiences to preserve the feeling of liveness. Such programs, in spite of an apparent diversity, continued to demonstrate the self-conscious ability of radio to create a sense of immediacy, which was to reappear in television's equally extensive range of programs. The *BBC Handbook* of 1938 explained that the Television Service provided something for everybody, including two and a half hours of "live" material as distinct from film, being available every weekday, together with an hour on Sunday evenings. The studio programs ranged from listed music, ballet, revue, art exhibitions, fashion parades, variety acts, and drama.[5]

The critical discourse constructed around the early function of television before the arrival of commercial television (ITV) in 1955 was therefore concerned with television as an aesthetic medium—but which aesthetic was "natural" to television? In 1970, Joan Bakewell, the presenter of the talk show *Late Night Line-Up* (1964–1972) on BBC television, was to comment that, "a whole theology was erected to justify television's existence as an autonomous art in terms of its liveness."[6] This "theology of liveness" becomes more intelligible by examining the nature of public broadcasting in the UK. Until 1955, television and radio were the exclusive preserve of the public service corporation of the BBC, and the Corporation existed as an institution that was self-regulating. It meant that it was uniquely privileged to be able to guide Britain's cultural development, and it championed the role.

During the immediate postwar period, the BBC promoted itself as an instrument (if not *the* instrument) of national cultural improvement on radio and television, and carried the competing claims of "information, education and entertainment" within its Royal Charter. Information was privileged not only out of a belief that television could not win in competition with cinema because of a temporary technological constraint such as the smallness of the nine-inch screen or the inability to record programs, but because the idea of broadcasting was rooted in a service ethos.

Despite these limitations, some early drama did explore television's formal possibilities. In 1938, experimentation took place in one of the earliest science fiction dramas on British television. Karel Capek's *R.U.R. (Rossum's Universal Robots)* was an adaptation of a popular play for television, which had been critically regarded in similar ways to modernist texts such as Yevgeny Zamiatin's *We* (1921), Fritz Lang's film *Metropolis* (1926), and Aldous Huxley's *Brave New World* (1932). Though popular, *R.U.R.* can be classified less as science fiction with a populist generic appeal—inherited to some extent, as we have seen in chapter 1, from the chapter play—than as an example of high culture. Moreover, when adapted and produced for television by Jan Bussell, it did not simply use television as the eye at the keyhole, but deployed the professional expertise of a creative medium.

Reviewed in the *Times* on February 14, 1938, *R.U.R.* was praised as a "good choice for television." "Rossum's Universal Robots are a fit subject for the television screen," the *Times* argued, because the television screen could provide a clearer image of the appearance of individual characters through the use of close-up. The close-up can be thought of as a special technique of the studio and continues to form an important element of the idiom of television presentation. Bussell himself commented that "television is unsuccessful with crowds, panoramas, spectacles, but it reveals the souls of men better than any other medium."[7] Handled like this, the appeal of television was the deployment of the close-up and the opportunity to exercise the attention, examination, and response of the viewer. Having been adapted for television, *R.U.R.* was produced for the audience at home rather than the theatrical audience. However, the final scene proved to be problematic:

> a sentimental last scene, in which two robots fall in love—a robot Adam and Eve—which seemed a mistake on the stage, and an even greater mistake in the intimate atmosphere of television.[8]

In the play, the robots invade the factory and shout "Finished" as they kill each man. But it was the ability of television to get up close and reveal the emotion of the final sentimental scene that jarred with the *Times* reviewer. He believed that it distracted from the earlier moment of horror when the robots had usurped the power of men. The close-up, to be used correctly, had to be judiciously exercised.

The close-up allowed for the examination or analysis of a scene, but it also provided other audience pleasures. It formed a strategy of intimacy and an empathetic use of the camera, which Jason Jacobs describes as "the revelation and display of the character's inner feelings and emotions, affected by a close-up style of multicamera studio production."[9] The use of the close-up was to achieve a strict fealty to the actor and a type of dramatic acting that depended on the idea that the distance from the camera was analogous to the psychological penetration of the character. Television drama based on close observation was balanced with sympathy for the characters in order to penetrate under the surface and reveal the hidden springs of their thoughts. However, it was the intimacy of the small screen that would lead many British filmmakers to dismiss television later as a medium that was only capable of producing a stripped-down image consisting mostly of close-up shots or "talking heads" rather than a "spectacle."[10]

The postwar improvement of telerecording—the recording of a live television program by filming it from a monitor, called kinescope in the United States—where it had become a tool for the purpose of repeats and syndicated sales to overseas markets—also raised possibilities of new production methods. However, it did not seriously challenge the belief, in Britain, that television was essentially a live medium. Large centralized complexes such as BBC Television Centre, completed in 1960, were products of the era of live television in which television drama consisted of the continuous performances shot in "real time" inside a studio. Yet, increasingly, program makers began to exploit the possibilities that recording offered, and the aesthetics of television drama began to change.

The figures for the output of television drama are revealing. By 1956, ten years after the postwar resumption of television services in Britain, drama had increased to a total of 225 plays. Of these, 116 had been specially written for television and 62 were adaptations of stage plays.[11] Adventure-story serials and series had become an established feature on Saturday nights.

Meanwhile, technical limitations in the form of fixed-focus cameras and a lack of recording facilities were also being resolved. The cameras had originally contained an Emitron tube, manufactured by the Emitron Television Company, with a 6.5-inch lens of a maximum aperture of f.3 which gave a subtended angle between 28 and 25 degrees.[12] Nigel Kneale, who penned the groundbreaking science fiction series *Quatermass* for the BBC (1953–1959), wrote about the swift development of television technology. The first series was shot live in 1953 and used prewar Emitron cameras, tracking on bicycle wheels. They had fixed lenses and a viewfinder that offered an image that was upside down and, because of its general ineffectiveness, was referred to as a "watch-the-birdie" viewfinder.[13] By 1955, and the second series of *Quatermass*, there was a well-equipped studio and filmed inserts were possible. Finally, when the third series was produced in 1958–1959, there was console lighting to control multiple lights at once, and recording using videotape.[14]

When Michael Barry became Head of Drama in 1952, succeeding Val Gielgud, he established the Drama Script Section, with a script supervisor under his direct control whose brief was to find new writing. Within a year the script supervisor reported "very encouraging" statistics, including that 12 out of 107 dramas transmitted over the past twelve months had been new works written specially for television (against hardly any in the previous five years).[15] Under Barry's leadership, this increased to 256 new works for television in the twelve months ending March 1960.[16] Barry's aim was to replace "dead wood" with new producers from film and theater.[17] The implication was that television on the same basis of regard, esteem, and concern was no longer a pastime between supper and bed, but could be placed with the novel and the theater.

Barry also spent his entire first year's budget for commissioning new scripts (£250) on one man: a young short-story writer called Nigel Kneale. Within a year Kneale had written the *Quatermass Experiment*, a six-part science fiction serial. The *Quatermass* science fiction series written by Kneale and directed by Rudolph Cartier in the 1950s was a sensational success because it was different from so much else on television. According to John Caughie, it marked a moment of original, as distinct from preexisting drama, adapted from the theater or novel, in the early history of British television. *Quatermass* was able to make use of filmed inserts to produce a hybrid dramatic form that combined the immediacy of the liveness of the studio with the action of filmed or recorded footage and signaled the expanded function of television. The first serial, *The Quatermass Experiment* was made in 1953; this was followed by *Quatermass II* in 1955, and finally *Quatermass and the Pit* in 1958–1959. *Quatermass*, as a representation of the fantastic, was able to appeal to audiences in ways that the "intimate" style of drama had not done. Television was no longer following a logic of remaining a transmitting medium but became one that, like cinema before it, could focus on telling stories by producing images to build a tight causal narrative chain of events.

It is as well to remember that the production of drama in early television was not only a question of adapting the scripts to the size of different shots, and types of camera movement, but also of capturing on television the nature of theater performance. Television relayed performance, and the live moment was transmitted as it happened in the same blocks of continuous time as theater.

> The studio was a performance space, full of technology and techniques of mediation, but placing these technologies and techniques at the service of an ideology of immediacy. The limitations placed on *mise-en-scène* and editing as expressive devices, the restraint of style which comes to be associated with "boring" naturalism, were not simply borne out of technological constraint or imaginative failure; rather they belonged to the logical aesthetic of a technology whose essence was conceived in terms of immediacy, relay and the "live."[18]

Jason Jacobs argues, however, that it was the arrival of Nigel Kneale and Rudolph Cartier at the BBC that challenged the "intimate drama." In talking about their early work, Kneale and Cartier placed particular emphasis on television's potential as a *visual* medium with a closer affinity to cinema than radio. Kneale was arguing for a style that could be used in the construction of a narrative through images rather than through the dialogue, and emphasized the role of the camera in visually constructing the drama. As Jacobs has argued in relation to Kneale and Cartier's adaptation of George Orwell's *1984* (BBC, 1954), their challenge to the idea that television was only capable of an intimate rendering of "the actor," helped to reject the notion that television was incapable of a more expansive aesthetic that could produce interesting images to tell stories.[19] Philip Purser, the television critic, would later explain about Cartier:

> I think he was terribly important because he was the first man to expand television, to make it big, to get out of the little sets and the little three or four shot confrontations, he really had unfettered ambition. He did it in the studio right up to and including *Quatermass and the Pit* in which he destroyed half of London. He did use a lot of film, he was a brilliant director of film—nobody ever matched the studio production so perfectly, and nobody ever got such instant quick quality which he got. He used a lot, more than the average director was allowed to use. But he only used it when he had to. Ahead of Sidney Newman he had this depth, he went inside the drama, instead of just viewing it as a series of boxes. He had enormous ambition. . . . The bigger and more difficult it was, he did it.[20]

The notion of television as a "cozy" medium was anathema to Cartier and Kneale, but it was the logical conclusion to a set of assumptions that depended on tropes about nearness in order to define television as a medium. A way of escaping the intimate and cozy was to explore different genres that demanded imaginative visualization of fictional worlds. As Joy Leman notes: "Science fiction as a genre in literature, film and television offers the possibility of moving beyond the dominant narrative constraints of realism and naturalism in exploring political ideas, visions of an alternative reality and dramas of fantasy."[21] The *Quatermass* stories of the 1950s had two primary and related functions: first, to explore the stylistic possibilities of television drama; and second, to expand the thematic range of television drama through generic form.

What was being discovered in the 1950s, according to Kneale, was that there was no single technique, no purist method that was "natural" to television. However, the pressures encountered within the electronic studio were also due to the problem of working within the constraint of the limited spaces available to live drama. Irene Shubik, who was to work on *Out of this World* (1962) and *Out of the Unknown* (1965–1971), explains:

> I'd hate to direct in a studio. In fact I'm beginning to hate working in a studio altogether, mainly because of the time pressure which, I find, is unbearable. There's a perpetual anxiety that you've got three hours in which to record a play and if you don't get through in three hours there's some poor director going mad, and you've got to be there telling him "For Christ's sake, get on with it."[22]

Ted Kotcheff, a Canadian, who had been working in television for ABC (ITV) with Sydney Newman on *Armchair Theatre*, from 1958 to 1960, was also frustrated by the technical difficulties of working in TV and aspired to direct for the cinema:

> You have no absolute control of [television's] picture quality. Your lighting effects which you took great trouble over, may go out of the window once the show is transmitted, because someone has their brilliance up too high, or contrast down too low . . . so you're working in a pictorial medium where you run the risk of losing 50 percent of your picture quality! . . . Adding up all the points, you find that ultimately working in TV is like sculpting in snow, it's all gone in the morning sun and everyone's forgotten it was ever there. This is why every TV director wants to get out into films.[23]

Nevertheless, Kotcheff was conscious that despite these problems, working in television afforded a greater freedom of action because the time pressure allowed for "mistakes to be made;" a situation that permitted experimentation.[24] Asked whether he preferred to work in TV or cinema, Kotcheff was to explain that "TV is very exciting and helps you to develop your ideas, and there's a great freedom of action once you're working in the studio working with the cameras because people can't interfere."[25]

It was the segregation of film operations at BBC Ealing from "live" television at BBC White City and Lime Grove that led to methods, at Ealing, that relied on those devised within the film industry, augmented by some new approaches.[26] Economics and availability of equipment were the factors that influenced production. Whereas feature film production had relied on few, but large and often lavish, studios, television film production required smaller studios that could be used flexibly. A feature film might be a year in preparation and months in production, whereas a television play could go from inception to screen in less than three months.

None of the programs in the *Quatermass* trilogy had a connection with its predecessor besides its eponymous hero, portrayed by a different actor each time: Reginald Tate in the first series, John Robinson in the second, and André Morell in the third. Kneale later explained that the original *Quatermass*, with its moon rockets, space suits, and alien monsters, "was a send-up of science fiction; a lot of it was meant as comedy."[27] By the second series, however, people were taking the series seriously. The third *Quatermass* series, written by Nigel Kneale and produced by Rudolph Cartier, was broadcast at peak viewing time across six Mondays from December 1958 to February 1959.

The final *Quatermass* series, coming at the end of the 1950s, was made during the beginning of a shift in science fiction as a genre. Three years later, the sci-fi writer J. G. Ballard would remark on the changes appearing in science fiction:

> I think science fiction should turn its back on space, on interstellar travel, extra-terrestrial life forms, galactic wars and the overlap of these ideas that spreads across the margins of nine-tenths of magazine s-f. It is these, whether they realise it or not, that s-f readers are so bored with now, and which are beginning to look increasingly out-dated.[28]

Sputnik, launched in October 1957, and the accelerating space race between the United States and Soviet Union continued to fascinate fans of science fiction. But there was a growing disenchantment with "nuts and bolts" science fiction, associated with rocket ships and the conquest of space. This disenchantment can be understood as a trend by some writers toward *speculative* fiction, replacing the older definitions and expanding the textual borders of science fiction. By the decade's end in 1968, science fiction writer Brian Aldiss—a contributor to the British progressive sci-fi magazine *New Worlds*, edited by the flamboyant Michael Moorcock—would produce a retrospective anthology of space travel stories whose last chapter, he believed, should be called "The Stars My Detestation."[29] The space race, as a

demonstration of the vocabulary of space hardware and the NASA-controlled narrative of the conquest of space emanating from Cape Kennedy, had failed to address the strangeness of the only planet known to be inhabited: Earth.

The thematic complexity of *Quatermass and the Pit* is therefore due to its being an exploration of the idea of the space race. Noticeable is the serial's pun on the word *race*, in its sense of ethnicity, as well as raising the possibility that the Earth was visited by alien astronauts—a space race in reverse—five million years ago. Presumptions about the human race's preeminence and its manifest destiny to occupy the planets are subverted by the discovery of the Martians on Earth. Although not a unique theme—Arthur C. Clarke had used it extensively in some of his early novels of the 1950s—it remained a novelty on television. Initially, Kneale uses "nuts and bolts" technology in *Quatermass and the Pit*, but the buried UFO quickly becomes a sign of the uncanny—a concept coined by Sigmund Freud in his 1919 essay "The Uncanny," described by Terry Castle as:

> An obsessional inventory of eerie fantasies, motifs, and effects, an itemised tropology of the weird. Doubles, dancing dolls and automata, waxwork figures, alter egos and "mirror" selves, spectral emanations, detached body parts ("a severed head, a hand out of the wrist, feet that dance by themselves"), the ghastly fantasy of being buried alive, omens, precognitions, déjà vu. . . . What makes them uncanny is precisely the way they subvert the distinction between the real and the phantasmic.[30]

The blurred distinction between the "real" and the "phantasmic" is encapsulated by the UFO. The unidentified flying object had been part of the repertoire of science fiction since the 1940s: specifically since June 24, 1947, when private pilot Kenneth Arnold reported airborne, saucer-like shapes while flying over the Cascade Mountains in Washington State.[31] The story quickly became a media mania as new stories encouraged more people to see the saucer-like craft themselves. It was Kneale's ingenuity to adapt the narrative of visiting alien craft to represent advanced technology able to traverse the huge empty distances of space, and connect it with an interest in the autochthonous and mystical. In many ways, the UFO phenomenon is a reminder of science fiction's relationship to earlier literary traditions, such as the Gothic and the ghost story. What are being represented in the final *Quatermass* series are the paranormal possibilities of the Fourth Dimension. *Quatermass and the Pit* offers a literal representation of outer space in the form of a spaceship in order to start a thematic expansion incorporating the evocation of recessed, occulted space. Finally, it is the machine apparatus of television that becomes the method that will be used to not only represent but analyze an alien psyche that has lain undisturbed for millions of years.

The series starts with footage, shot on 35 mm film, of construction workers laboring at a building site in Hobbs Lane, London. The "excellent definition" of the pictures was remarked on by a test audience questioned later by the BBC, although the opening footage, shot on film, possesses a different materiality from the telerecorded footage recorded in the studio.[32] The workers' discovery of a skull starts a train of events that leads to a team of archaeologists discovering what appears, initially, to be a bomb deposited in the soft London clay where the skulls are buried. Set in the west London district of Knightsbridge, the scenes were in fact filmed in Cheapside, then in the grip of major postwar reconstruction designed to repair the devastation of the Blitz. The appearance of a black man as a construction worker is a timely reminder of the arrival of new labor from the Commonwealth, and the unease felt by many white viewers about "colored" immigration.[33] These themes of race and conflict, which form the thematic basis of this sci-fi drama, are prefigured by the black worker, although he is less superstitious than the other (white) construction workers, reversing the usual expectations about race and belief in primitive, supernatural forces.

The scenes on film were either shot on location, at Ealing Film Studios' studio 2, or on the Ealing Studios lot. The basic studio was the BBC's Riverside Studios in Hammersmith, where the program was broadcast live and recorded on film. At the time, Riverside Studios was one of the best-equipped and more suitable for a major project such as *Quatermass and the Pit* than the BBC's studios at Lime Grove, where *Doctor Who* would later be recorded. *Quatermass* is a demonstration of how early television was able to remediate not only film but also live performance. The invention of telerecording would mean that a television image could be recorded on film, processed, and if necessary projected immediately as a filmic remediation of television. However, television was also transforming the experience of viewing film, as well as live performance, into a televisual discourse. Moving pictures had been on the cinema screen, and the cinema, with its mammoth images, had had to satisfy an auditorium. Television was to remediate the audience's relationship to film by its broadcasting into the home.

This is, in many ways, reminiscent of Raymond Williams's argument that television is a technology, as well as a coexisting cultural form.[34] He believed that television as a specific technology had remediated multiple cultural forms, including theater, radio, and film, and remade them as television. We have seen how television inherited from radio a diverse range of program forms—music, revue, variety, drama—but these were organized to form a continuous schedule. Yet, Williams's belief that this creates a unified and singular cultural form fails to fully notice the heterogeneous form of television. Williams had sought to rescue television by analyzing its form in order to confront the commercial values and state interests that he saw as dominating television. But to understand television as the outcome of specific interests is to suggest that the discourse on television is more singular than in fact it is. For example, the adaptation of the aesthetic of the live camera for television as a piece of drama in *Quatermass and the Pit* is handled very differently from the social realism of the 1960s, exemplified by Ken Loach's *Cathy Come Home* (1966) shot on 16 mm film. Rudolph Cartier's directorial style in *Quatermass* is too cinematic to be compared to those later dramas of the 1960s but, nonetheless, raises issues about liveness and realism within television.

INNOVATION AND AUTHORSHIP ON TELEVISION

It is significant that Cartier, like other famous creators of early television programming, has slipped into obscurity and been subsequently rediscovered. In this sense, television is indeed an ephemeral medium and presents less of a canon of works than is found in the cinema. Writing in 1990, Julian Petley and Linda Myles considered why Cartier's work was so little known, concluding that the reason was "the continuing ascendancy of what one might call the Loach/Garnett/Sydney Newman School of television drama in critical histories."[35] This effect had been noticed as far back as the early 1970s, when Cartier's work had been recognized as different from that of Loach, as well as from Jack Gold's, a director known for working in the British Realist tradition.[36] The UFA-quality of chiaroscuro in Cartier's productions was markedly different from the naturalism visible in the realistic drama of the 1960s that became in the 1970s a "gritty realism." This form of realism would dominate much of filmed television drama and heavily influence the last *Quatermass* series made in 1979 for Thames Television.[37]

Cartier, who brought with him a particular filmic sensibility that had been informed by Continental drama, described working in television as a "foot in both camps." He argued that although both film and television told stories visually and with the aid of sound, that was where the similarity ended. In an interview published in September 1958, he discussed the ontological differences between cinema and television, remarking on the small size of the television screen (between twelve and twenty inches in the mid-to-late 1950s) and the belief that the television would be watched in a darkened living room.[38] Cartier made the point that, owing to the British system of broadcasting (in the late 1950s) on the prewar standard of 405 lines, the image would not be sharp enough to see the facial expressions of the characters. It was therefore necessary to bring the story close to the viewer and Cartier was able to express his enthusiasm for television and his belief that, unlike film, it had a "hypnotic" power, due to the isolation of the viewer at home. Watching a program alone in a darkened living room was, he believed, unlike sitting in a darkened cinema with others. "When the viewer was watching these "horrific" TV productions of mine [*Quatermass I and II*], he was—I like to think—completely in my power, and accepted the somewhat far-fetched implications of the plot . . . while, in the cinema, there was usually a titter or false laugh."[39] As a consequence of these differences, the impact of television was such that it could be said to emotionally *move* the audience more than film.

One of the setbacks of film directing is that the director has to shoot the story "out of continuity" in order to accommodate set construction and other studio considerations. On the other hand, the television director within the electronic studio works in strict sequence of events, giving the actors a sense of events building up in the performance. However, because television is a live broadcast, the performance must be rehearsed for two to three weeks before it is transmitted, in order to avoid mistakes being made during broadcast. As a consequence, on television the actors will deliver "good performances," which might be extrapolated to mean that the performance will be more authentic and, as a further consequence, the audience will empathize with it more. Cartier was recalled as a director who was a stickler for precision and detail in a performance:

> Each director of course has their own approach and style, even when trained by the same organization. . . . The good television director cherishes his cast, bolsters its morale and calms its astonishing insecurities . . . however . . . Rudolph Cartier [was] like those directors in the early Hollywood films, who wore britches . . . always sort of cracking whips. He was ruthless.[40]

Cartier's work was multifaceted, and included a high percentage of literary adaptations. In fact, his foray into science fiction was marginal to his output, though not necessarily marginal to his career, because successful science fiction had enabled him to obtain larger budgets to produce his literary and continental adaptations. Controller of Television Programs Cecil McGivern had recommended a bonus for Cartier due to the success of the original *Quatermass* serial and Michael Barry would observe the following year: "Mr. Cartier . . . brings an adventurous imagination in television and is keenly ambitious about what may be done with it as a means of creative expression. Like the rest of us he has made his mistakes but they have been the mistakes of 'over-reaching' and not mediocrity."[41] Notwithstanding the success of *Quatermass*, Cartier's staple contributions continued to be adaptations of sophisticated and usually canonical European drama, novels, and operas. Cartier had the ability to use both an intimate style and, when the demand arose, a more expressive style. Crucially, it was Cartier's call for the BBC to adopt "a new approach, a whole new spirit, rather than endlessly televising classics like Dickens or familiar London stage-plays," which was an encouragement for new narrative material, including such unfamiliar and outré material as science fiction. *Quatermass and the Pit* adapts the intimate style to include cinematic elements such as crowd scenes, spectacle, chiaroscuro lighting, and a complex mise-en-scène imbued with symbolic significance. Its formal hybridity reveals a complex system of adaptation of often opposing styles.

Despite his varied work, Cartier, in part due to his émigré status, can be regarded as an auteur within BBC television. His work is somewhat unified by his repeated use of close-ups and an emphasis on performance rather than expensive production design. The emphasis on performance suggests again an intimate style on television but, at the same time, Cartier could place more emphasis on the cinematography over the dialogue. It was the ease with which he could harness and begin to transcend the limitations of television, while maintaining grandiose ambitions more in keeping with the larger budgets of feature film that became personal qualities that afforded him a career at the BBC. Cartier's strength was to be able to deploy his artistry inside an industrialized medium.

Cartier was dismissive of recorded film or television films: "In my opinion . . . it is only live programs that comprise real television, in the best sense of the word." On the other hand, he was to claim:

> This will not mean the end of "live" cinema. Quite the contrary. People will always go to the pictures, as no home entertainment can compare with the thrill of expectancy when sitting in the midst of a thousand people, waiting for the lights to slowly dim, the curtains to part and the familiar card The British Board of Film Censors . . . to appear before the screen.[42]

Cartier's enthusiasm for live television belies the fact that he remained in "both camps." The idea of "live cinema" is curious and contradictory, but can be understood as the expectancy of watching film "when sitting in the midst of a thousand people" compared to the solitariness of watching television.[43] Although, this method of understanding the ontology of cinema and television and the audience's relationship to it appears to be too elastic to offer any final meaning, it is useful to reflect on Cartier's claims about the essential differences between film and television, while not fully accepting his conclusions.

Born in Austria before World War I, Cartier's original name had been Kacsar. He had been trained as an architect but had had ambitions to become a director of opera before becoming fired with an enthusiasm for films and becoming a scriptwriter for a series of films being made in Weimar Germany. He had been taught by Max Reinhardt, the famous impresario, who believed a script should be similar to a musical score, and interpreted by a director in the way a musician should interpret a musical score. At the German film studio UFA, which had made the science fiction film *Frau im Mond* (1929), Cartier's closest colleagues were Emeric Pressburger and Billy Wilder. In 1935, when he was unable to continue to live in Germany any longer, Cartier traveled to England. After the war, he became a BBC staff producer, a position encompassing the roles of director and scriptwriter, and showed an interest in productions with supernatural themes: *The Duybbuk*,[44] based on the original Jewish play, and *Portrait of Peter Perowne*,[45] whose life-after-death plot makes for an interesting comparison with the film *A Matter of Life and Death* (US: *Stairway to Heaven*), which had been scripted by his colleague from Germany, Emeric Pressburger. Cartier was to say about himself, "I became something of an expert on fantasy and science fiction."[46] He was to go on to direct two contributions to the BBC science fiction series *Out of the Unknown* ("Level Seven" and "The Naked Sun"), and one for the *Late at Night Horror* series ("The Triumph of Death"). However, although Cartier was to refer to *The Quatermass Experiment* as "a science fiction serial," he wrote privately that he was "anxious to lift [it] above the level of strip-cartoons."[47] Ultimately, the program was listed in the *Radio Times* as "a thriller in six parts."

Cartier's sensibility was born out of his formative experience of being a Central European, and by working in Weimar Germany. He had been given the freedom at the BBC to choose source material that included modernist and absurdist playwrights, including writers such as Brecht, Sartre, and Anouilh. It was also the use of film, especially, as Julian Petley points out, in the shooting of locations that extended the range of stylistic choices available within *Quatermass and the Pit*. Crucially, however, Jason Jacobs argues that the use of film inserts in television drama was not the adoption of a cinematic style. Rather it was to provide a more heterogeneous style that challenged dominant notions about television as being an intimate medium. It is important to remember that intimacy and immediacy, though often used in the same breath about television, can in fact be different. A large-scale event may not be "intimate," but if it is shot in real time, it can be presented as "immediate." The use of film inserts, Jacobs argues, expanded the space of the production.[48] The expansion of space can be extrapolated as offering extra-diegetic "color." By shooting outside and filming four minutes or so of exterior shots, it was possible to add what had been missing from the black-and-white television image. The use of film encouraged stylistic changes that broke from the low-key realistic mode of representation that had hitherto existed. The monochrome austerity of British television was to be challenged by Cartier's more "colorful" Central European sensibility.

Although "color" can be equated with a geographical exoticism, it can also be allied with new physical and psychological spaces to create a discourse of the implausible and absurd, in opposition to other institutional discourses set by the BBC that favored, at times, realism. In the 1960s, the doyen of art and cultural commentary on British television—Huw Wheldon, who anchored the TV arts show *Monitor* (1958–1965)—was to explain that, "I'll welcome colour when it has reached its full potentiality. In the mean-

time, *limitation is the friend of style*."[49] Wheldon's remark can be understood to link the issue of style back to the time-based nature of television programs: television's original ephemerality. Television, conceived in these terms, existed as a medium that, despite using similar production techniques as the cinema, deliberately avoided the cinema's expressive imagery. *Quatermass* was to bring not only the world into the home, but the worlds of inner space that writers such as Ballard had declared to be of more interest than nuts and bolts space rocket engineering. Nigel Kneale would explain his own thoughts about the science fiction genre in two interviews:

> I don't like the term "science-fiction," but if we're going to bandy it about, it could be applied just as well to the world we live in. The form is appropriate, if taken seriously. And that is the way I do take it. I try to give those stories some relevance to what is round about us today. The last one, for instance, was a race-hatred fable that broke through to an encouragingly large and intelligent audience. On the technical side, it went about as far as possible towards exploding the "intimacy" fallacy. Huge sets, long shots, crowd scenes were the order of the day. One critic remarked: "Not only does it sweep away detachment, but it obliterates also the feeling of being a solitary spectator; one reacts to it with an enlarged response as a member of a communal experience."[50]

> I always feel that the most interesting strange thing was to have an ordinary setting. I like the blending of the ordinary and the extraordinary, the funny and the horrific.[51]

QUATERMASS AND THE PIT

In the opening scenes of *Quatermass and the Pit*, workmen excavating a building site in the Knightsbridge district of London unearth the fossilized bones of three skeletons: apparently primitive humanoids, but with enlarged and oddly shaped skulls. "Knightsbridge ape-men," to give them the racy sobriquet attached by onlookers, are a bizarre intrusion into the fashionable district, once degraded by German bombing but now (in the late 1950s) in the midst of economic renewal. Initially, *Quatermass* is set in a contemporary, recognizable world and serves the narrative in two ways: first, it fulfils Kneale's desire for the drama to be "televisual"; second, televisual realism is contrasted to the "cinematic" exoticism of the spaceship.

As well as television form and style, the opening episode deals with another theme: the cultural status of television itself and its ability to take complex ideas and otherwise successfully represent them. Richard Hoggart had published *The Uses of Literacy* in 1957 and Raymond Williams's *Culture and Society* appeared the following year. Both books were part of a wider reappraisal of older definitions of culture as to what could be considered to be of value and what could be labeled as tawdry. Nigel Kneale would have been aware of these arguments when, in a 1959 interview, he explained that "The intelligent viewer must be won and kept, otherwise TV will just go down the drain. . . . This will then lead to automation; electronic computers that turn out a beautiful plot, complete with dialogue and period costume."[52]

The call to arms to raise the status of television by addressing the "intelligent audience" is due to that medium's centrality to cultural development and, surprisingly for a piece of popular entertainment, thematically it occupies a significant portion of the first episode. It is first amusingly dealt with in an early scene during an encounter between the younger purveyors of the tabloid press, the *Evening Gazette*, and a much older writer from the respectable academic journal, the *Palaeontologist*. The scene reveals the avoidance of fact for journalistic hyperbole by the popular press. However, the attack on television itself, because its definition as a *popular* medium may constrain what it represents, is more startling. Due to its popularity, television can be understood as a device that encourages an active citizenship. But this is critiqued during a wry observation of the technique of *vox populi* (the voice of the people) and in-built assumptions about the value of public opinion. The use of the vox pop and inclusion of public opinion is often a characteristic of news programs and documentaries, but the claim that such shows can inform the viewer is mocked.

A middle-aged, well-spoken female journalist is standing at the site where the ape-men have been discovered. She holds a hand-held stick microphone. The microphone is designed for commentaries in noisy, outdoor surroundings and excludes extraneous sounds. The scene is being filmed, but not broadcast electronically, using an outside broadcast (OB) unit. Until the advent of relatively lightweight TV equipment, most electronic programming originated in television studios, and film crews with 16 mm cameras covered most news programming and high-mobility situations. A snaking queue of members of the public is waiting to see the ape-men. The journalist begins by addressing a young couple, whose dress and performance marks them out as a working-class "Teddy boy" and "Teddy girl."[53] She asks them "what do they think about this business about the missing link?" The man replies, "I don't know . . . it's all right if they want it." Flustered by the reply, the journalist asks the young woman her opinion. She responds that she "likes it." When pressed about "liking the missing link," the woman becomes defensive. The journalist tries again with another member of the public and receives another illogical answer. Finally, she announces that she has caught sight of someone whose opinion is "really worth something"—Doctor Matthew Roney, a paleontologist overseeing the exhumation of the skeletons.

The scene is highly revealing because it provides a clue about Kneale's attitude to television, and possibly Cartier's. It should be remembered that the BBC had already criticized Cartier because "his métier does not lie in what might be loosely termed the more popular field of entertainment."[54]

However, Kneale believed that *Quatermass and the Pit* "broke through to an encouragingly large and intelligent audience." The implication is that those producing the show were concerned that it would not break through because television is a medium that, conceived as being mass entertainment, will only appeal to a domestic audience, which is either distracted or passive, but not active. To conceive of cultural power negatively, especially in regard to popular programming on television, may refute the need to develop a positive analysis of cultural value in a medium that is itself a form of mass communication. However, television as a relay medium was adapted to produce a richer text that incorporated a sophisticated narrative and visual style. The fear, nevertheless, remained that the audience would desire popular, undemanding drama: unchallenging science fiction with its ray guns and alien monsters. The BBC had itself been complicit in this desire and had broadcast a warning to build interest in the forthcoming serial: "By now the nature of the *Quatermass* serial is well-known to viewers and we therefore leave it to those of you with a nervous disposition to make your own decisions as to whether you view this program." The same warning was to go out half an hour before each episode was broadcast.

It was clear what the audience wanted from research undertaken by the BBC.

> A substantial group maintain[ed] that the story had got off to a disappointingly slow start, and that the bloodcurdling events foreshadowed in the trailer had not, alas, come to pass. "I thought it opened badly," a Telegraphist declared. "It took almost half an hour to get going. Also I was looking forward to seeing some spine-chilling drama, but this was not forthcoming." There were, in addition, a number of complaints to the effect that the episode was "rather disjointed" and "didn't seem to hang together," that it was "muddled and confusing," "bewildering" as well as "unexciting." It may well be that a few, at least, of the critics had failed to grasp the implications of the so-called bomb's position relative to the fossil remains and the absence of an "ingress cavity," an Insurance Representative, for one, commenting: "Unexploded bombs turn up in very peculiar places and I couldn't see why any excitement or suspense should be caused by one being under the excavations." . . . The majority, in contrast, were either looking forward to a livelier turn of events in succeeding editions, or (and this was much more frequently the case) were waiting with bated breath for further fascinating disclosures (and the more gruesome the better, several exclaimed) in what promised to be "an absorbing piece of science fiction."[55]

The public service ethos practiced by the BBC had led to a suspicion of cultural elitism. Several scenes in the first episode, easily overlooked, represent the venerable institution in the form of a well-spoken and educated woman confronted by the illogicality of the "masses." The BBC's challenge in the late 1950s—developing television as a mass medium, no longer broadcasting to a privileged, metropolitan middle class—was becoming acute after the arrival of ITV, the instigators of "people's TV," in September 1955.[56] The new broadcaster had made it clear in the editorial of the first issue of the *TV Times* what the failings of the BBC monopoly were considered to be:

> Television is at last given the real freedom of the air. The event is comparable with the abolition of the law that kept motor-cars chugging sedately behind a man carrying a red flag.
>
> Now it's the "go" signal, the green light for TV too—with no brake on enterprise or imagination.
>
> So far, television in this country has been a monopoly restricted by limited finance, and often, or so it seemed, restricted by a lofty attitude towards the wishes of viewers by those in control.
>
> That situation now undergoes a great and dramatic change. Viewers will no longer have to accept what is deemed best for them. They will be able to pick and choose.
>
> And the new Independent Television planners aim at giving viewers what viewers want—at the time viewers want it.[57]

Such a demotic appeal to the *populi* helps to explain some of the ambiguity that must have been felt by Cartier, the cultured Central European, and Kneale, the writer of ideas, about working for the mass medium of television, as it developed in the 1950s. Nevertheless, earlier *Quatermass* science fiction series had enjoyed enormous public support, and with the arrival of ITV and the prospect of it as a rival channel, Cecil McGivern, the controller of BBC television programs, argued that the new competition meant that the corporation should be producing many similar dramas. He wrote that "[Nigel Kneale] . . . gave us the serial, *The Quatermass Experiment*. Had competitive television been in existence then, we would have killed it every Saturday night while that serial lasted. We are going to need many more *Quatermass Experiment* programs and series."[58] One reason for the commissioning of a third science fiction series that adapted the same character to a new series was therefore a desire to prove that the BBC was less of a culturally conservative organization than had been claimed by the supporters of ITV. However, the tension between television as a mass medium and the sci-fi genre that can be regarded as "ideas-driven" is both visible in *Quatermass and the Pit*, and inscribed into it. *Variety* may have confirmed Kneale's worst fears:

> Though it was confidently predicted that the recent "Quatermass and the Pit" six-episode thriller would, for the first time since commercial television started operations in Britain in 1955, put BBC-TV at the top of the ratings, confidence and predictions have in fact been shattered.
>
> In particular, it was thought that the last instalment, aired on January 26, would zoom way out in front of every rival program. Press interest had built to a zenith, and the immensity of public enthusiasm was suggested by a motion

> put at one local council that business shouldn't start until after the "Quatermass" transmission had ended. . . .
>
> Yet on the Nielsen charts the cliffhanger didn't figure anywhere in the top 10 . . . in those homes where both BBC-TV and commercial programs can be received. . . . Still more watched "Keep It in the Family."[59]

Therefore, the finale of the *Quatermass* series, described as "an avuncular homily strangely out of place," becomes more comprehensible: the didacticism is serving a purpose at odds with the promise of entertainment.

In the next few scenes of episode one, we see Professor Quatermass at a meeting with another key character of the series: Colonel Breen, head of the Experimental Rocket Group. Breen is planning to build bases on the moon capable of launching nuclear missiles at the Earth, and thus providing a deterrent against attack from hostile nations. For Quatermass, the decision to build military bases on the lunar surface is a watershed moment. Humanity will either take its jealousies and hatreds into space, or it will abandon them and seek a brighter future. Such sentiments are dismissed as naive idealism by Breen and the other participants at a high-level meeting, which is chaired by a government minister. Quatermass's exasperated response is that the militarization of space will encourage other countries to build their own lunar bases. A representative of the older military establishment, an elderly general, is quick to remark "So there is a race on."

The general's comment alludes to both the nuclear arms race and the space race, but an additional play on the word *race* becomes increasingly evident as the series develops. The ape-men buried beneath the streets of Knightsbridge are evidence of an attempt by arthropod Martians to escape their own dying planet and colonize the Earth. Kidnapped by Martians millions of years earlier, they were anatomically altered to give them Martian-like brains—psi powers, a "group mind," and a deeply engrained hatred of any Outsider—in a body adapted to life on Earth. Returned to Earth aboard a fleet of spacecraft, the altered ape-men were intended to act as proxies for their Martian creators, dominating local species and conquering the planet. The object that the workers at the Knightsbridge site first take for an unexploded bomb is revealed to be one such spacecraft, whose crash landing millions of years earlier was fatal to its mixed crew of arthropod Martians and ape-men. The Martians' instinct for purifying the race and promotion of a closed society based on rigid hierarchies is used to draw parallels with the military mind, represented by Colonel Breen.

Dramatic lighting is deployed in *Quatermass*, including a bold use of color: black and white. Julian Petley and Lynda Myles have remarked on the almost UFA-like quality of chiaroscuro lighting in much of Cartier's work. Chiaroscuro had also been striking in Cartier's earlier production of George Orwell's *1984*, which is a masterful adaption of the original novel. Cartier was capable of using expressive lighting in an era when within the television set it was characteristically "flat." The use of three to five electronic cameras in the studio would mean that flat lighting would be used in order to be able to shoot a subject from all angles at the same time without any apparent discontinuity in lighting and background. Flat lighting would also avoid shadows cast by boom microphones during multicamera recording inside the studio. At the same time, a director was aware that adding lighting effects was difficult because the viewer at home might adjust the contrast on their television set if the picture was too dark or too bright.[60] Moreover, many television engineers in the studio would strongly disapprove of brilliant whites and solid nonreflective blacks, especially if they were in large quantities, because they were of the opinion that the electronic signal of the TV system would not cope with highlights and lowlights. Alan Hume, a cameraman at the time, comments:

> What, for example, looks more ludicrous than a person creeping about with a torch when there is so much light that you can see everything as clearly as in daylight? This sort of result is usually caused by the cameraman *thinking he must play safe, he probably being of the opinion that the T.V. system will not cope with dark, unlighted areas.*[61]

Using this production logic, the electronic technology of the studio could be made subordinate to the script and made invisible in order to disguise the technical nature of television. For example additional lighting might be added to avoid nonreflective blacks. However, as Hume was to point out:

> There are surely times . . . one is getting good pictures which are interesting, or even exciting; the lighting is adding to the atmosphere and the suspense, and plays a very important part in helping the director and scriptwriter to put over their story and to make it more interesting.[62]

Cartier's approach within the TV studio was on the whole more innovative.

The use of stylized lighting can be inappropriate in a multicamera studio. Yet there are scenes in *Quatermass and the Pit* that combine heightened drama and emotional realism because of the use of chiaroscuro lights. This occurs in the first scene in episode two when Professor Quatermass and a policeman walk to a derelict house built over the site of the crashed spacecraft and abandoned more than twenty years earlier due to its poltergeist-like noises (later revealed to be connected to the presence of the Martian ship). If only a short scene, it is eerily effective by deploying the codes of expressionism and the use of heavily contrasting areas of light and shadow. The visual style illustrates the growing suspense and paranoia of the script as the normal, stable world becomes chaotic, while beneath their feet the spacecraft is being exhumed. We watch Quatermass approaching the abandoned house and the urban decrepitude is reminiscent of the "chaos world" that is familiar in the film noir thriller. Recorded using the process of telerecording, the image lacks sharpness, but is remarkable because the filming is treated as a photographic event in which lighting, camera, actors, and background are completely integrated. The scene cuts from Quatermass outside the house to an interior mid-shot of the door opening. Quatermass's shadow on the wall of the hall suggests, momentarily, an ethereally menacing figure waiting inside as the scientist enters.

By holding a torch, the scene fails to be realistic—and recalling the quote from Alan Hume, there is too much light from the torch—but it continues to demand attention. The policeman who has accompanied the professor enters holding another torch and explains to Quatermass that the house was abandoned in a "ghost scare." In the gloomy atmosphere of the house, the thin line between the natural and the supernatural blurs. The policeman becomes increasingly agitated as Quatermass questions him about the events of twenty years ago, which he witnessed as a small boy. The British bobby, lionized in films such as *The Blue Lamp* (1949), starring Jack Warner as the redoubtable George Dixon, crumples in this scene. The director's power is to erode our natural skepticism about the occult. The shift of the series from science-fiction to the Gothic is accomplished primarily by the expressionistic mode of lighting. However, it is the cutting between the multiple cameras in the studio that allows lighting to be set up separately for each area of the house in ways that are similar to the separate setups of filmmaking.

Nevertheless, the expansive style is matched by an intimate style in the following scene, set in the house of a medium, which feels disappointing because of its flat lighting and the limited mobility of the camera. The scene is shot using the common technique of frontal acting in a multicamera studio: the actors face the camera the entire time. In the scene, the woman who lives next door to the abandoned house describes the events of the "ghost scare." Yet the scene fails to dramatize the account until the end when the camera is moved on a crane to stop on a close-up of the medium. Such contrasting scenes demonstrate how although Cartier was to admire the skills and techniques borrowed from the feature film, he was able to exploit as well as be constrained by the techniques of shooting in a multicamera studio. *Quatermass and the Pit* was made before flexible editing would have made it possible to record short scenes like a film. Nevertheless, Cartier's achievement was to overcome the obstacles of working under pressure on a complex production, and deploy a high degree of stylization.

Although the basic studio used for the production was the BBC's Riverside Studios, the Martian spaceship was mainly shot inside Studio 2 at Ealing Film Studios, which could better handle scenes that included water or special effects. The set depicting the excavation site around the spaceship, for example, was designed so that the walls and foreman's hut could be raised upward, creating the illusion the pit had become deeper. A climactic scene set inside the spaceship brings the set design to the fore. Quatermass and his team penetrate a sealed door marked with a pentacle, revealing chambers with a honeycomb geometry that prefigures the discovery of the insect-like Martians who have been trapped in a self-made catacomb for eons since the crash. The light source in the scene is a powerful lamp held by a young army officer at the far end of the spaceship. Another man, also a soldier, faces the camera and is much closer. This technique of having the actors facing the camera was common in low-budget film-making, where, due to the need to shoot quickly, separate setups are avoided. However, here it is a more "legitimate" form of filming, although the frontal acting toward the fourth wall again reminds the audience of the nature of the electronic studio, and the set's lack of any real depth.

The use of music by Trevor Duncan and electronic music composed by the fledgling BBC Radiophonic Workshop, which had been set up in 1958, adds to the growing sense of expectation about the discovery of the Martians. Dick Mills, at the time a technical assistant at the Workshop, would produce the weird oscillations that suggest the psychic power of an alien civilization not completely dead yet. Compressed air forced into pipes resembling cables was used to make the spaceship come alive, with the cables rising off the ground in a demonstration of unleashed psychic powers. A night scene depicting the Martians' lingering psychic effects taking possession of a workman called Sladden demonstrates again how Cartier was to negotiate a foot in both camps. He was to argue:

> Television, I might explain, has not yet discovered how to transmit night scenes, as the electronic tube does not agree with low-key lighting. It suppresses all middle and dark tones, and the result is a greyish blob without any graduation, interrupted by chalky white highlights, also without graduation. We hope this may be remedied in the not-too-distant future by a better tube.[63]

Yet Cartier was to prove that a photographic style could meet the demands of the production, which did not have to meet the same criteria as an engineering test card. Lighting is once again an effective element of the mise-en-scène as Quatermass and his team try and drill a hole through a bulkhead in the spaceship to discover what lies on the other side. Removing the bulkhead, the interior is lit to reveal gelatinous and scintillating webs left by the Martians. When Sladden is possessed by the alien forces, which have been unleashed by the removal of the bulkhead, the entire incident is recorded using imaginative lighting. The scene is intercut between the POV of Roney's assistant, Barbara Judd, and a low angle camera on Sladden. The camera creates identification with Judd as she watches Sladden pursued by an unearthly force that sends debris flying into the air as he makes his way out of the pit. The scene again demonstrates Cartier's mastery of a mobile camera and lighting inside the studio.

Quatermass and the Pit is not a demonstration of an opposition to the "service aesthetic," which has been identified by John Caughie, Jason Jacobs, and others. Rather, it is evidence of the complexity of television as a medium and how far it could be adapted to go against the dominant paradigm of an intimate medium or a dramatic form that was essentially "naturalistic." It was possible for a television auteur to exploit generic verisimilitude. By orchestrating a multitude of memorable and complex images, authorial power counters the concept of televisual flow as it was initially suggested by Raymond Williams.[64] According to this theory, flow was a process in which broadcast TV tended to minimize and deface the differences between programs. For Williams, flow was a feature of TV that compromised and altered both the separate texts that TV had produced and the texts that it had appropriated to its own use, for example, feature films. Instead, his model was one of self-contained texts that, because they appeared in a flow, had had their distinctive characteristics reduced or eliminated. On the other hand, the use of mise-en-scène in *Quatermass* appears to belong to a televisual canon, a set of dramatic masterpieces, in which particularity stands out from the general. John Ellis has explained that "Every age has its own canon of great texts and one criterion for inclusion in any canon of classics is simply that a text should be amendable to use beyond the confines of the historical context in which it was generated."[65] The selection of a canon has been problematic for television, but a canon can be attractive because it concentrates on specific texts as a method of discriminating between the materials that exist in the television flow.

The method of production within the electronic studio has meant the idea of a television auteur is difficult to justify. According to David Bordwell, the art film "foregrounds the *author* as a formal component, the overriding intelligence organizing the film for our comprehension."[66] A television drama that aspired for authorial expressivity was limited by the conditions of working inside a multicamera studio, but this has to be balanced against the claim by Cartier that he had the television audience in his "power."

Graham Murdock has pointed out that analysis of the role of the writer in television drama has tended to start either from the notion of "authentic" and "creative" authorship or from the notion of the craftsman constrained by the restriction of a ratings-orientated industry. A term that is useful here is *wordsmith*, which implies an ability to work to order and fit the requirements of writing for a regulated form like television.[67] For Murdoch, there was a "primary" sector of the cultural industries devoted to producing material that was designed to maximize audiences.[68] On the other hand, a "secondary" sector used an ideology of authorship that accorded the writer expressive autonomy. By the time television got under way in the 1950s, the major ideological and institutional divisions within television were already firmly established, and television drama was obliged to accommodate them.

Anthony Smith's appendix to the Annan Report (1977) had this to say about the role of creativity within an industrialized medium such as television:

> Every radio and television producer has his own relationship with his employing organization and his colleagues; it is based on his professional record and past attainments. His situation within the fairly complex processes of broadcasting is always a special one, in that he is working as a kind of "author" directly inside a manufacturing and distributing process. He is one of a relatively small number of creative professionals performing his work among people with a variety of different skills, many of which would apply equally well in other industrial processes; he, however, is dependent on the machinery and the administration and the scores of people who supply these, to perform a fundamentally intellectual and creative function.[69]

These were some of the limitations on the director and the writer. Moreover, studio time is not allocated by the needs of the production, but instead is dictated by the fact that studios have to function efficiently as a form of production line. Andrew Pixley explains that when *Quatermass* was in preproduction

> Cartier was doing his utmost to get more time and facilities. Claiming that Kneale's complex story merited a longer than usual reprise, the producer asked to expand his slot by five minutes to 35 minutes—suggesting that the show could begin at 8.25pm, taking up a five minute buffer normally allowed with the preceding program, *The Phil Silvers Show*. Cartier was given the green light for this just over a week later on Tuesday 25. By Monday 8 December, most of the episodes were rescheduled to 8.00pm, with BBC Television's line-up re-arranged around the star serial.[70]

As we have seen, Cecil McGivern, the controller of BBC television programs, had argued that the new competition from ITV would compel the corporation to produce more dramas like *Quatermass*.[71] But the idea that the new *Quatermass* show would be a "star serial" suggests not simply the importance of the author as the unique enunciator of the text and its overall meaning, but how television's expanded function has always been dependent on the economics and operation of the institution in charge of production. Ordinarily, the show must be tailored to fit into the time as has been scheduled for it. But Cartier was able to exert pressure to be afforded extra time, and this ambition for the serial did not stop there, despite the assembly line nature of production. Within long working days, there is little slack in television production, and everyone works within the tight parameters of budget and time. For example, the filmed component of *Quatermass* was shot from the morning of Monday, November 24, 1958, to Friday, December 12. However, the dedication of the production team can be appreciated by the fact they worked fourteen-hour days. Moreover, broadcasting the first live episode on December 22 would mark the start of another hectic schedule before the second episode was ready for broadcast.

As has already been mentioned, the length of time available in the studio did not permit individual setups as a rule and one of the consequences of not using separate camera setups was that television shot in the studio continued to use flat lighting. At the same time, the lighting in a studio is never tangible until the lamps are positioned to illuminate the actors and set. Nevertheless, within tight and rigid deadlines for each scene there is little opportunity to make major changes. The temptation is accordingly to light safely by ensuring the shots are adequately exposed for each camera (either outside or within the multicam studio). This means allowing light to enter from all directions, eliminating shadowed areas, but creating a featureless, bland look. The risk of using expressionist lighting in *Quatermass* was to become too ambitious and to end up with producing ill-matched shots and excessive shadows that appeared uncontrolled. Cartier would be aware that the lighting crew were BBC employees who wished to continue to work for the corporation. An innovative lighting effect that fails attracts criticism and affects the reputation of the man who was responsible. In an organization run on draconian commercial lines, the pressure not to innovate is strong, but it was to the credit of the BBC that experiments in the use of lighting were permitted. Cartier was a director who possessed dramatic vision, but also understood television's limitations as much as its possibilities in pursuit of an expressive style within the organization of the BBC.

Quatermass and the Pit posits that the Martians' attempt to colonize Earth was at least partially successful: that the ape-men they imbued with their ruthless and clannish nature are the ancestors of much (but not all) of humankind. The final episode of the series concludes with Professor Quatermass addressing the camera, delivering a warning to humanity about the dangers of its own instinctual desire for conflict. The ending is an affirmation of the power of television as a medium that because of its live ontology speaks directly to its viewers. The speech is addressed didactically to an audience:

> If another of these [spaceships] should ever be found, we are armed with knowledge. But we also have knowledge of ourselves and of the ancient destructive urges in us that grow more deadly as our populations increase and approach in size and complexity those of ancient Mars. Every war crisis, witch hunt, race riot, and purge is a reminder and warning. We are the Martians. If we cannot control the inheritance within us, this will be their second dead planet![72]

Although didactic, the speech is also highly dramatized. A studio camera and a boom microphone are made visible as a method of creating a sense of immediacy. The setting appears to be a broadcast lecture on the events that have taken place, but it can be better understood as a commentary on the series with the intention of elucidating its purpose.[73] The cuts between the television cameras end by focusing attention on Quatermass using a close-up. The conclusion to the series was not to the liking of Michael Barry, the head of Drama, who complained that it was "an extremely pompous and in some ways inconclusive end."[74] The ending can be read as awkward because it is anticlimactic (the 1967 Hammer film adaptation would not make this mistake), but it demonstrates the heterogeneous nature of television, its expanded role as educator as well as provider of entertainment. It can also be understood as an acknowledgment by Kneale that television was less about communicating straight facts than about enlivening those facts. Tables of figures about race hatred and its threat would be difficult to take in, but a narrative built up of personalities, especially the figure of Quatermass, is easier. Using audience research from the BBC and internal evidence from episode one, it can be shown that it satisfied reservations by Kneale that the series should offer a clear understanding of the ideas it contained rather than end with an anticlimactic tag or epilogue scene; the conventional ending in a contemporaneous series such as the popular police procedural show *Dragnet*.

CONCLUSION

Both *Quatermass and the Pit* and Disney's "Man in Space" dramatize the "space race"—except, as we have seen, Kneale was able to bring an entirely new interpretation to the notion of "race." The competitive spirit of the space race had been the uplifting of national spirit and a source of pride in the United States, but due to its demand for the coordinated use of massive industrial resources it was symbiotically linked to the military and the Cold War. In "Man in Space" there is, at times, a disturbing continuity between the German *vergeltungswaffen* (vengeance weapon) program and the American military embrace of the new rocket technology, which is fetishized in "Man in Space." *Quatermass and the Pit* is best understood as adapting the televising of the space race, its iconography and its values, occurring in the United States:

> By the end of the decade space travel was a common feature both in American cinema and on American television and, as a result, a well-honed topic with an established iconography: multi-stage rockets belching smoke on lift-off, weightless humans, heavily cratered lunar landscapes and intrepid astronauts journeying into the unknown. The aura of success that Americans garnered

> from this regular diet of narratives and images of brave and fearless space travel was rudely shaken by the Soviets getting into space first.[75]

Each rocket launch was a propaganda coup to demonstrate the superiority of American science and technology over the Soviet Union.[76] Mike Allen has argued persuasively that the space race is an event that can be understood as one of many events from the Civil Rights Movement to the Vietnam War that television was able to present as the medium became an integral part of US society. As television became the nexus for communication in the United States,[77] British science fiction diverged from its American cousin. The United States, which possessed the industrial resources of a superpower, had became engaged, despite some initial setbacks, in a titanic race to build rockets that would launch men into space, as well as carry nuclear warheads—the development of the intercontinental ballistic missile (ICBM). Britain, marginalized from the space race and confronted with the loss of its world role as an imperial power, did not have the same confidence.

> Thus the notion of the British in space was not only a depressing impossibility but an outright absurdity. Some of the basic differences in character between British and US science fiction TV may be traceable to this: where archetypal US series like *Star Trek* often confronted the future with a sense of gung-ho optimism, British equivalents were more prone to view it with pessimism, anxiety.[78]

This difference accords with the decline in British economic and military capability after World War II. However, philosophical differences can be traced to less tangible sources, as well as imperial and economic decline. For Brian Stableford, the allure of the extraordinary has become a category of fiction, which he labels as "British scientific romance." *Quatermass and the Pit* poses problems to the science fiction aficionado. The combination of occult themes—mediumship, poltergeist activity, and cabbalistic symbols, sit awkwardly with the idea of invading Martians and advanced techniques such as "atomic surgery" performed on the ape-men to boost their intelligence. Yet, within this thematic mishmash it finds its antecedents in the seminal figure of science fiction: H. G. Wells. According to Stableford, the scientific romance is the product of a peculiarly English sensibility, but it held risks for its writer:

> The great success of Wells's scientific romances was not considered to constitute a redemption of scientific romance *per se*, but rather a testimony to his artistry in being able to write stories that were worthy of applause *in spite of* their nature. . . . For a few brief years, though, scientific romance *was* fashionable as the new periodicals went through their experimental phrase. That period began in 1893, reached its height around 1898, and had petered out by 1905. While it lasted, a great many magazine writers were permitted a measure

> of self-indulgence without stigma—a licence to play with some exciting new ideative toys without too much embarrassment.[79]

Part of the repertoire of ideas in *Quatermass and the Pit* is anthropologist Robert Ardrey's theory of man as "killer ape," now unfashionable, but immensely popular at the time. The idea was repeated in the opening of *2001: A Space Odyssey* (1968), which appeared less than a decade after *Quatermass*, and is also the basis of the *Planet of the Apes* (1968). The first two texts posit the idea of a Darwinism that is not Earth-bound, but an interplanetary selection of the fittest, visible in the chilling scene in *Quatermass* depicting the Martian hives being "cleansed" of the weakest members of the Martian race. These "ideative toys" enable a commensurate style to exist. Cultural experimentation and the sense of risk-taking were also possible because, within the parameters existing for British television, broadcasting had not become as commodified as it had in the United States by the end of the 1950s. If the service aesthetic defined aesthetic ambition,

> the informal working relations, the culture of the BBC canteen, the "try it and see" attitudes, carried something of the "wizard prang" about it, an extension of church hall dramatics which suggests a space which was not fully institutionalized. In many ways, this period parallels the experimentalism and lack of standardisation characteristic of "primitive" cinema as described by [David] Bordwell and [Nöel] Burch. The significance of the late 1950s, with the arrival of competition and the technology of recording, was to install an increasing professionalism in the place of the enthusiastic amateur, and to begin the process of the institutionalisation of the mode of production of television.[80]

Experimentalism in television suggests a binary: the enthusiastic amateur and the television professional. However, the allure of the extraordinary helps to transcend this binary. *Quatermass* was able to benefit from the larger budgets assigned to it and had been allocated a significantly larger budget than the earlier *Quatermass* serials—£2,560 per episode.[81] It would be a mistake to suggest that Kneale or Cartier were amateurs in the true sense of the word, but Kneale, in particular, had begun working in television when the medium had been in its formative stages and lacked the strict hierarchies that existed later.

The fate of Rudolph Cartier is, perhaps, instructive here. After Michael Barry left, he was eventually replaced by Sydney Newman, whose populist approach to drama had been evident in the encouragement of social realism on *Armchair Theatre*. Newman has been criticized because he had "industrialised the making of television drama."[82] Cartier now found himself assigned to productions such as half-hour episodes of *Maigret* (1961 and 1963) and later *Z-Cars* (1963). By 1964, the annual report on Cartier's performance revealed the tensions he felt inside the BBC and the frustrations of his colleagues:

> Rudy Cartier is a law unto himself. He is extremely courteous and bends an attentive ear to all that is said to him—and does exactly what he wants to do. He is a first-class publicist and is second to none in getting facilities within the Corporation to work for him. Keep him off "art" and he will turn in a respectable and extremely popular show.[83]

By 1965, Cartier's desire to develop an expansive style appeared less innovative and his annual report commented that his recent productions, while "nearly always hav[ing] the popular touch . . . fail to penetrate in depth and reveal little more than that which is on the printed page of the script."[84] By 1966, it was observed that Cartier's "enthusiasm for and interest in the television medium . . . [was] still as genuine and refreshing as ever," but his directorial style was no longer thought to be "in the fashion of the moment."[85] The style of directing that was thought to be fashionable and to be encouraged was not simply social realism but, as always, the adaptation of a hybrid style, which remained fluid. This was never more the case than in the science fiction anthology *Out of the Unknown*.

NOTES

1. Mackie, "Six Hundred Hours a Week."
2. Buscombe, "All Bark and No Bite," 200.
3. Gielgud, "Policy and Problems of Broadcast Drama," 23.
4. Madden, "Television," 225.
5. *BBC Handbook* (1938), 40.
6. Bakewell and Garnham, *New Priesthood*, 14.
7. Bussell, *The Art of Television*, 106.
8. "Televised Drama: Karel Capek's *R.U.R.*," *Times*, February 14, 1938.
9. Jacobs, *Intimate Screen*, 8.
10. "A Turnip-Head's Guide to the British Cinema," written and directed by Alan Parker (Thames Television, first broadcast March 12, 1986). Ken Russell argued that British cinema and Channel Four's films are basically B movies, even if some of them are very good B movies.
11. *BBC Handbook* (1958), 86.
12. Baily, *The Television Annual for 1950/51*, 36.
13. Kneale, "Not Quite So Intimate."
14. Ibid.
15. Robin Wade, *Where the Difference Began: Some Developments in Script and Script Sections in BBC Television, 1936–1974*, BBC, Television Script Unit (internal document detailing history of the Drama Script Section), 1975, 6. Document is in the collections of the British Film Library.
16. Statistic quoted by Wilson, "Introduction," 14.
17. Barry, *From the Palace to the Grove*, 187.
18. Caughie, *Television Drama*, 44
19. Jacobs, *Intimate Screen*, 110.
20. Philip Purser, interview with Jason Jacobs, June 29, 1992 in ibid., 131.

21. Leman, "Wise Scientists and Female Androids," 108–9.
22. Shubik, "Producers and Directors," 87.
23. Porter and Wicking, "The Making of *Life at the Top*," 230.
24. Wheatley, "And Now for Your Sunday Night Experimental Drama."
25. Porter and Wicking, "The Making of Life at the Top," 230.
26. Ealing Film Studios were bought by the BBC in 1955.
27. "Intelligent Viewers Must Be Captured," *Television Mail*, November 13, 1959, 16.
28. Ballard, "Which Way to Inner Space?" 2–3.
29. A play on the title of Alfred Bester's 1956 science fiction novel *The Stars My Destination*.
30. Castle, *The Female Thermometer*, 4–5.
31. Of course space ships existed before then and adorned the copies of the pulps. However, as a disk-like shape and as a semi-mystical concept, the UFO finds its origins in Arnold's observation.
32. Audience Research Report for *Quatermass and the Pit*, episode 1, "The Halfmen." BBC WAC T5/2,302/1.
33. The most notorious race riot in England until the 1980s had taken place in August and September 1958 at Notting Hill, London.
34. Williams, *Television: Technology and Cultural Form*.
35. Myles and Petley, "Rudolph Cartier," 129; Cartier has since become better known, thanks to work such as Jacobs's *The Intimate Screen*, which looks at his 1954 adaptation of *1984*.
36. Pattison, "The Man to Beat," 17–18.
37. Sexton, "The Origin of Gritty Realism."
38. Cartier, "A Foot in Both Camps."
39. Ibid., 10.
40. Richard Hewett, "Acting for Auntie: From Studio Realism to Location Realism in BBC Television Drama."
41. Michael Barry, "Rudolph Cartier Annual Report 1954," April 6, 1954.
42. Cartier, "A Foot in Both Camps," 31.
43. Ibid.
44. Originally broadcast October 21, 1952.
45. Originally broadcast December 2, 1952.
46. Quoted in Myles and Petley, "Rudolph Cartier."
47. Rudolph Cartier to C. Moodie, BBC WAC T5/418.
48. Jacobs, *Intimate Screen*, 129.
49. Huw Wheldon quoted in John Pearson, "Carleton Greene: The Man with a Growing Empire," *Sunday Times Preview of Colour TV*, 1962, 9. Italics added.
50. Kneale, "Not Quite So Intimate."
51. Quoted in Petley, "The Manxman," 91.
52. "Intelligent Viewers Must Be Captured," *Television Mail*, November 13, 1959, 16.
53. The Teddy boy/girl subculture which emerged in the mid-1950s was distinguished by stylish neo-Edwardian clothes (boxy jackets over tapered trousers and skirts, silk brocade waistcoats, exotic neckties), elaborately styled hair, and (after 1956) a devotion to rock and roll.
54. January 1, 1957, BBC WAC L 1/2177/1.
55. Audience Research Report for *Quatermass and the Pit*, episode 1: "The Halfmen." BBC WAC T5/2,302/1.
56. Also the fear of "Americanization," which is outside the scope of this book.
57. "Editorial," *TV Times*, September 22–October 1, 1955, 3.
58. Cecil McGivern, "The Future: Script Section" (memo), November 10, 1954. BBC WAC T31/141/1.
59. *Variety*, February 18, 1959.
60. Ted Kotcheff makes this point when discussing the problems of working in the television studio. See "The Making of Life at the Top."
61. Hume, "Filming for Colour Television Series," 5. Italics added.
62. Ibid., 5.

63. Cartier, "A Foot in Both Camps."
64. Williams, *Television*.
65. Ellis, "Is It Possible to Construct a Canon of Television Programs?," 25.
66. Bordwell, "Art Cinema as a Mode of Practice," 59.
67. Murdock, "Authorship and Organisation."
68. Ibid.
69. Annan, "The Relationship of Management with Creative Staff."
70. Pixley, "The Martian Inheritance," 39.
71. McGivern, "The Future: Script Section."
72. *Quatermass and the Pit*, episode 6, first broadcast January 26, 1959.
73. This form of didacticism was popular on television at the time. Oxford University historian A. J. P. Taylor was, at the time, presenting his own lectures on television, in which he would try to elucidate a complex topic from modern European history in only thirty minutes.
74. Quoted in Chapman, "Quatermass and the Origins of British Television Science Fiction," 38.
75. Allen, *Live from the Moon*, 15.
76. Ibid., 44.
77. Allen mentions that by 1963 television had overtaken newspapers as the primary means by which people accessed the daily news. See ibid., xiv.
78. Cook and Wright, *British Science Fiction Television*, 4.
79. Stableford, *Scientific Romance in Britain*, 127–28.
80. Caughie, *Television Drama*, 55.
81. Chapman, *British Science Fiction Television*, 36.
82. Taylor, *Days of Vision-Working*, 190.
83. Rudolph Cartier, Annual Report 1964, BBC WAC L1/2177.
84. Rudolph Cartier, Annual Report 1965, BBC WAC L1/2177.
85. Rudolph Cartier, Annual Report 1966, BBC WAC L1/2177.

Chapter Three

Brave New World

Out of the Unknown

PEOPLE'S TELEVISION, INNOVATION, AND EXPERIMENTATION

Despite the success of science fiction on the BBC, it had been neglected by the ITV companies: regional broadcasters, operating under a central regulatory body, established as alternatives to the BBC under the Television Act of 1954. The BBC had screened *Quatermass* during the 1950s and another drama serial in seven parts—*A for Andromeda*, written by the noted astronomer Professor Fred Hoyle, and scripted by BBC producer John Elliot—in 1961. Both shows were adult science fiction compared to the bug-eyed monster (BEM) type, which had been disdained as semiliterate pulp fiction. The trade journal *Television Mail* was happy to report that:

> *A for Andromeda*, the BBC's new science fiction serial started with a gripping first installment . . . with a literate script and with an authentic real world contemporary flavour. The use of fictional "actuality" news interviews, (remember the newsreel interviews in Citizen Kane?) heightened the credibility and tension of the story. . . . The production with its fluent transitions from studio to film, the re-creation of a ten pin bowling alley in this way was particularly neat, its swift changes of locale gave the whole show a sweep which was difficult to resist. The control centre sets were completely convincing, with the possible exception of the telescope back-projection plates, and gave a feeling of elbow room, a pleasant change from the often rather claustrophobic effect of small cluttered-up studio production. . . . The cardinal virtue of the series remains its assumption that the viewer is on nodding terms at least with galaxies, NATO, computers and radio sources and has the imagination to grasp the significance of a radio message from outer space.[1]

Meanwhile, ABC-TV (Associated British Corporation Television) had won one of the ITV regional broadcasting franchises in 1955. ABC-TV was bright and brash and was later perceived to appeal more to the young than some of the other ITV companies. Earlier, its managing director, Howard Thomas, had hosted the BBC radio show, the *Brains Trust* (1941–1961), which had been an arena for lively, intellectual debate. Subsequently, on becoming managing director of ABC-TV, he had helped to make it one of the "big four" ITV companies. The desire for a science fiction show with strong thematic possibilities was initiated by Thomas before being passed on to his head of Drama at ABC-TV, Sydney Newman.

Sydney Newman was one of many Canadians who had been offered employment in British television in the 1950s and 1960s. Newman had worked in New York, before being offered the position of head of Drama at ABC-TV in 1958, largely because of his achievements on a series of *General Motors Theatre* productions, purchased by the BBC. This had brought him to the attention of Howard Thomas. Newman is credited with two accomplishments at ABC-TV that reflected his North American background: a populist approach to drama that used a form of social realism on the celebrated anthology series *Armchair Theatre* (1956–1974), and the innovation and encouragement given to the production of popular series such as *Police Surgeon* (1960) starring Ian Hendry and, when that series stopped, the pairing of Hendry with Patrick Macnee in the original series of *The Avengers* (1961).

Newman soon passed the project of developing a new science fiction series for ITV onto Irene Shubik, who had been story editor of *Armchair Theatre*, and had already worked as a documentary scriptwriter in America and Britain. The idea of a science fiction drama had always been a pet project of hers. In some ways, the new science fiction series *Out of this World* (1962) was to be a spin-off of *Armchair Theatre*, which as a groundbreaking drama anthology had sought social realism to mirror life at the end of the 1950s. The working relationship Shubik was to enjoy with Newman meant that she had comparative autonomy to do as she wished on scripts:

> We had a monthly meeting with Sydney Newman and presented him with all the people we'd seen and ideas we'd like to commission. We'd discuss them all and he'd say "OK. Go ahead on this—go ahead on that." And he never saw the script again until it was in a shape that we considered was absolutely produceable. The minute you got a script, you know, you talked to the director and the designer. It was all little conferences. Unfortunately, you just about never get this sort of early discussion here [BBC] because it's so big.[2]

Within ABC-TV there existed a freedom to overcome institutionalized attitudes about "how things were done." Shubik as a woman, and Newman as a Canadian, were able to bring with them a greater informality and the possibility of experimentation that within parts of the BBC, shaped by patriarchal and Oxbridge attitudes, made things difficult. Asked how she looked back on her *Armchair Theatre* days, Shubik responded:

> Everyone says they've never had that feeling again of a whole unit of people who had the same ideas and approach to production. . . . When you got a script, the designer came in and said "Wouldn't it be more interesting if we set this scene here and changed that set to shoot it more imaginatively?" . . . It's like doing things by computer here [BBC]. You just feed the script in and out comes the system's answer. . . . The good things about the BBC are that you have a tremendous number of technical resources and facilities available—like all the film units which you don't have in most commercial companies—and also you are more or less left alone. [3]

Out of this World was produced at ABC-TV for the summer of 1962 by Leonard White, who had been a producer for *Armchair Theatre*, and each episode of the series was introduced by Boris Karloff. There was wariness or even hostility to science fiction that was considered to be "pulp" or an example of a BEM. Instead, Shubik argued that science fiction on television should look to writers like Isaac Asimov, Arthur Clarke, Clifford Simak, and Kurt Vonnegut, who had some of the most original and philosophical ideas to be found in fiction writing. The disdain given to pulp science fiction echoes the importance of the service ethic prevalent at the BBC, but also, to a degree, at ITV. Rather than a vulgar type of entertainment, the corporation had, as a general attitude, wished to define culture in its educational and aesthetic sense. British television therefore sought to a certain extent to equip the viewer with the means to acquire a better understanding of "high" culture.

At the same time, the service ideal attracted criticism that it was elitist, and *Armchair Theatre* had sought to resist the high cultural associations of experimentalism. *R.U.R.*, adapted in 1938 and 1948, had in the latter adaptation styled the robots as classical mythical figures rather than metal men. The symbolism of their controlled, precise movements, while at the same time the performers remained visibly human, meant that there was less room for the literalism of the original robots and an opportunity for further abstract interpretation. But such abstractions and the arrival of ITV had led to general accusations of elitism against British drama. Meanwhile, the desire for popular drama on ITV acknowledged the need to sell "soap powder," due to its method of funding from advertising. Consequently, *Armchair Theatre* preferred a modified naturalism or type of social realism. Although the demands of the "sponsor" on British television were far less intrusive than they had been in 1950s American television—advertising was strictly controlled by ITV's official regulator, the Independent Television Authority—ITV sought to be "people's TV." Because there will always be an innate conservatism in a large, mainstream audience, within science fiction there existed a tension between the demands of an ideas-driven genre and what a general audience might accept on the television.

Yet science fiction as a television genre eschewing the ordinary or everyday would provide not only the "allure of the extraordinary," but be a site for experimentation on screen. It was the representation of novel ideas that would demand stylistic and production innovation. However, Leonard White was averse to applying the term *experimentation* to his productions. For him, *experimental* was to be replaced by alternative terms such as *innovation* and *difference*, as had happened in *Armchair Theatre*. ABC wished to adapt science fiction short stories by some of the best writers within the genre, but as an alternative to the high-brow elitism of the BBC. The television audience would be a popular audience for a mass medium and, in the era of the BBC/ITV duopoly, it would aim at the general audience rather than a specifically science fiction audience.

The search for suitable stories for adaptation to television posed a problem. Shubik consulted with Edward "Ted" John Carnell (1912–1972). Carnell had entered fandom as a young man and by 1938 was editing the *Journal of the British Interplanetary Society*, which included other science fiction enthusiasts, such as Arthur C. Clarke. In July 1946, Carnell was to produce a small pulp entitled *New Worlds*, and after a rocky start it finally became a digest-sized, monthly publication. *New Worlds* was to become Britain's most successful science fiction periodical in the 1960s. Carnell was also the literary agent for many British science fiction writers and a friend of John Wyndham, Brian Aldiss, and Arthur Clarke; he also promoted new work by such rising writers as J. G. Ballard and Kenneth Bulmer. He was to play a key role in helping to find suitable stories after realizing that

> there are thousands of short science fiction stories, but most of them have settings that were impossible to realise on the scale and budget dictated by television. . . . Short stories have too little to sustain narrative drive for fifty-plus minutes whilst novels have too much. . . . And the duration of a show can be problematic—thirty minutes may lack depth whilst sixty may drag.[4]

The science fiction writer and editor Alfred Bester, author of *The Demolished Man*, winner of the inaugural Hugo award in 1953, would echo this sentiment in a letter to Shubik:

> I don't envy your job. I was story editor on a half-hour sf show some years ago, and found it easier to write originals than ferret out suitable material. Somehow most science fiction does not lend itself to TV adaptation.[5]

One of the reasons Shubik had been successful in shortening the planned running time of the episodes from its original seventy-five minutes to sixty was because of the worry that the stories could not sustain a more lengthy adaptation. The essential elements of the original story, it was decided, could be adequately narrated in the revised running time. Bester's output of short stories had been prodigious, but in a letter to Shubik dated February 16, he had explained the problems he had had adapting half-hour fiction a few years earlier when editing *Tom Corbett: Space Cadet*. It was, he insisted, far easier to write original stories for television.[6] On the topic of adapting science fiction to television, Shubik would conclude:

> Unfortunately, there is no such thing as an actual science fiction script to be found anywhere. They did one series called *The Outer Limits* recently but that was film and 50 minutes in length, made in Hollywood, and probably intended for overseas sale. Apparently, there has never been a series on American television like the one we plan. The only near approach is, of course, *The Twilight Zone* which is much shorter in length.[7]

One consequence of having limited budgets and the constraints of recording on videotape in the electronic studio was that *Out of This World* was noted for its lack of physical action, although "for my money this has been all to the good," according to one critic.[8] The fourth episode, "Imposter,"[9] was judged to have been a failure "which had possessed a lot of movement and [in] a fair number of scenes, budgetary limitations in sets was painfully apparent, thus spoiling the potential appeal of a good play." The same critic would go on to say, "Correspondingly, those shows with a minimum amount of action came over well because the directors had sharper terms of reference and emphasized dramatically the intimate nature of the story."[10] Unfortunately, "Imposter" does not exist today, due to the policy at that time of wiping or erasing videotapes so that they could be used again. With the exception of only one episode, "Little Lost Robot," all thirteen episodes of *Out of This World* were wiped.[11]

The policy of wiping tapes was partly historical. When the BBC had reopened its television services after the war in 1946, "they virtually perpetuated a method of transmission on 405 lines, with the acquiescence of the Labour Government of the day. Excepting for BBC2 [it used 625 lines], British television has stayed on that standard like a truck on a narrow gauge railway."[12] The problem was the United States and Japan had adopted a technical standard of 525 lines and Europe had adopted 625 lines. This made it technically difficult and costly to make transfers from 405 lines to other video standards that could be exported. One unforeseen consequence of a "hurried" reopening of television services in 1946 would be that 16 mm or 35 mm film would have to be used if British programs were to be exported. "For exporting taped TV programs . . . transfers had to be made to film," wrote one commentator, although, "the quality of the results were (and still are) as unpredictable as the charms of trout fishing and women."[13] Unlike a program on video, a series that had been shot on film was to have clear export potential, but a program recorded on videotape had little or no value, and was wiped.

The high cost of videotape was another factor, which meant that it was unusual to archive programs, unless they had a continuing commercial value and might be exported, in which case, the videotape was copied onto 35 mm film. Film was a universal medium that could be exported to countries around the world that were using very divergent electronic standards in broadcasting—Britain used the Phase Alternating Line (PAL) system, and the Americans used the National Television System Committee (NTSC) standard. The differing standards meant that broadcast equipment was incompatible and tapes from one system could not be used on another system and vice versa.

One consequence has been that much of early television history is missing and cannot be reviewed. The ephemeral nature of television was noticed by practitioners at the time.[14] However, when *Out of this World* was transferred to the BBC and retitled *Out of the Unknown*, a few episodes survived. The departure of Sydney Newman to the BBC to become head of Drama meant that many from his original team at ABC-TV joined him there, including Irene Shubik. She transferred to the BBC in November 1963 as story editor of a science fiction anthology for the new channel, BBC-2.

By 1964, the BBC had won the right to broadcast a second television channel. It is probably misleading to suggest that there was a single causal reason for the setting up of BBC-2. Much ink has been spilled about the criticism of ITV's popular output after it began broadcasting and the outcome of the Pilkington Report (1962), which recommended a second channel for the BBC, on account of the quality of its programs compared to its commercial rival. However, Huw Wheldon, a leading figure at the BBC at the time and controller of programs by 1968 was to argue:

> I think that even had there been no ITV there certainly would have been a BBC2. The BBC situation is that if you're going to do a public service broadcasting job you have got to have two networks in order that you can provide choice, and in order that you take in the great minorities as well as the great majorities. One of the limitations is that you are unquestionably dealing with the large audience. . . . You are inescapably in the seven-figure audience league on both networks. . . . Two networks are essential. . . . The first basis of policy is to use two networks to broadcast what you think ought to be broadcast. . . . I like the idea of BBC being a channel of enthusiasts . . . a channel that cheerfully takes big risks.[15]

However, the claim that BBC-2 was interested solely in quality ignores the fact that its scheduling was designed to respond to BBC-1, and not vice versa. Nevertheless, for Wheldon the relationship between the channels was, as he put it, between BBC Lime and BBC Grove, each sharing the commitment of providing a public broadcasting service. For him, mediocrity was the real enemy.

Underlying the continuing desire for innovative drama at the BBC was an acute awareness of the importance of securing audiences for popular drama after the accusations of elitism and the ending of the BBC's monopoly in 1955. The tension between popular television and experimentation has been examined by John Hill, who argues that the BBC afforded creative possibilities for its most skilled and self-conscious practitioners, under benign management. There was the creative freedom to put into practice something exclusive to the medium. Such possibilities found their most concrete form in the Langham Group, formed in 1958, which unlike the populism of *Armchair Theatre*, was tasked to consider "the problem of experimental television programmes."[16] Yet the Langham Group was to be a failure, a cul-de-sac of endeavor that led nowhere. The television writer and critic Troy Kennedy Martin had dismissed the group's efforts, "[It was] an art set-up . . . propitiated on the altar of prestige," and did not believe that the group's ideas could be applied to "mass audience viewing."[17] Yet the group's attachment to visualization and recorded sound and stories that could not be told in conventional ways meant that it had the potential of developing possible forms for the treatment of fantasy. It was to "permit the testing of new techniques that would extend and refine the existing vocabulary of television drama . . . that break away from the inheritance of theatre and cinema, and from which eventually . . . will evolve something that is exclusive to the medium."[18] Here we have the ideation of television specificity, which opposed the use of film inserts and which regarded the electronic studio as the proper place to develop a television aesthetic. However, the Langham Group was not to survive long, as it competed with the gathering momentum for social realism in television drama at the end of the decade and into the 1960s.[19]

SERIOUS AND POPULAR DRAMA

The logic of filming longer film inserts on the newer type of increasingly portable 16 mm film cameras stemmed from a desire to include observational detail of the external world outside the studio. In order to make drama about contemporary life, television practitioners such as director Ken Loach and writer Tony Garnett sought to get out onto the streets, into the real world. The desire for greater "authenticity" in drama led to a tendency, beginning in the mid-1960s, to identify the studio with entertainment rather than "serious" types of drama. It was these opposing tendencies in British television, both related to the "nature" of television as an electronic medium, that were, in the 1960s moving British television drama (including science fiction) away from the intimate style and toward a greater diversity of forms.

The intensification of certain production tendencies that took place during the 1960s highlights both the possibilities and the limits of style in television. We have seen how, in *Quatermass and the Pit*, the desire was to break free from the trammels of "intimacy." Cartier was to bring an expressionist mode to the series, as well as a sense of scale—the use of crowds, for example—to suggest a greater realism. Moreover, Nigel Kneale was to say, "The uncovering of a space-ship from deep, real mud was shot from the full height of a Transatlantic crane—a session in a film [Ealing] studio that later added valuable realism to scenes in the electronic studio, where 'mud' means a sprinkling of peat over canvas and the small Mole-Richardson [a crane that permits the camera to hurtle around the studio] is the biggest crane there's normally room for."[20] Mole cranes have become considerably bigger and heavier than in Kneale's day, but the limits of the studio were clear. It is outside the scope of this book to discuss the development of the filmed science fiction series made by such people as Lew Grade and his Incorporated Television Company, and its development of a "celluloid television" culture shot entirely on film.[21] However, several of the episodes from the first series of *Out of the Unknown* help to illustrate the various production crosscurrents affecting the adaptation of short stories that were to form British television's brave new world of science fiction during the 1960s.

Such heterogeneity within production can be contrasted to a purely formal or aesthetic understanding of genre, including science fiction. A structuralist approach to genre can identify a form characterized by common features that reappear in different films. In this way, a genre is an object that has particular properties that can be categorized and studied. Tzvetan Todorov, whose work on the fantastic tale in the early 1970s became a scholarly landmark, makes clear how this approach works:

> The members of a genre share the inheritance of certain conditions, procedures and subjects and goals of composition. . . . In primary art each member of such a genre represents a study of these conditions, something I think of as bearing the responsibility of the inheritance. There is, on this picture, nothing one is tempted to call *the* features of a genre which all its members have in common.[22]

Similarly, Thomas Schatz has argued that genre can be studied as a formalized sign system whose rules have been assimilated through cultural consensus.[23] However, Schatz also explains that any critical analysis must be based firmly not only on an understanding of the genre but also the production system in which any individual genre film is made. Unfortunately, despite the caveat mentioned by Todorov that a formal or aesthetic approach to genre cannot be applied universally, but remains historical or diachronic, the lack of uniform criteria for genre delimitation has posed problems when thinking about genre on television. Jane Feuer makes the point in her essay in *Chan-*

nels of Discourse that genre analysis does not work as well as a paradigm for television as it has for literature or film due to the diversity of textual material available.[24] She also finds that television genres appear to have a greater tendency than older media to recombine across genre lines. Consequently, *Out of the Unknown*'s diversity of styles and narrative forms continued to guarantee meanings and pleasures for audiences, even if science fiction on television became increasingly an unstable marker of expectation and knowledge by the mid-1960s.

Sociocultural verisimilitude informs many British dramas in the period under consideration here, but it is necessary to recognize the range and multiplicity of forms not only on television but within a single series, and the possibility of reworking generic verisimilitude, extending and transforming it altogether as it appeared on television. Catherine Johnson, in her analysis of British telefantasy, has offered a useful schema (see table 3.1) to compare serious and popular drama on British television during the 1960s.[25]

These oppositions, however, are too reductive if accepted prima facie. To claim that ITV was capable of producing populist TV against the stuffy paternalism of the outmoded BBC falls into a trap created, in part, by those who campaigned most for independent television and the end of the BBC monopoly. It is necessary to think across, and beyond, these narrow binary oppositions when considering *Out of the Unknown*.

OUT OF THE UNKNOWN

In 1964, Shubik suggested a BBC version of her ABC show *Out of this World*, and later that year, Newman and Michael Bakewell, head of plays on television, commissioned her to become story editor of an anthology series of science fiction plays on the new channel, BBC-2. The type of drama was *modern* science fiction that avoided anything considered to be old-fashioned and that was more appropriate in an era characterized by Harold Wilson's "white heat of science and technology."[26] Shubik attempted to define what the series wished to offer:

Table 3.1. "Serious" Drama vs. "Popular" Drama

"Serious" Drama	*"Popular" Drama*
Social realist	Action-adventure/Generic
Deals with social concerns	Stylish and playful
Challenging and controversial	Entertaining and popular
Politically motivated	Commercially motivated
Primarily BBC	Primarily ITV

> To my mind Science Fiction, good Science Fiction, is a sort of modern mythology. Fables for our time. It is a mythology in which the background is that of a civilisation much more advanced technologically than our own, inhabiting a much wider universe. Our Science Fiction characters are Olympians in that they have so much greater scientific and technical resources than we do, but basically they remain human beings like us.
>
> You start out therefore, by supposing a future which is based on current scientific research and current conditions being carried to their logical or possible conclusion . . . over-population, space research etc. You then ask how will men react under these conditions, what sort of men will evolve. The fable writer can point any moral. If we allow such and such a condition to develop what will it be like? he asks. He can then be satirical, fantastical or pseudo-scientific as he wishes.[27]

The Science Fiction Writers of America bulletin of January 1966 suggested what a script for *Out of the Unknown* should contain:

> BBC Television wants stories for its new series of TV dramas. Production limitations rule out monsters, underwater adventures; the dramas are taped in the studio and are essentially live performances. . . . BBC likes symbolic or satiric stories—those with something to say about today. Endings should be conclusive, not the sort of thing that leaves the viewer hanging.[28]

Shubik had originally wanted the stories in the first series to be themed consecutively. This, she believed, could be achieved by establishing a specific period in time, and using the convention of costume, makeup, and setting throughout. Eventually, a diverse set of stories was produced using the most desirable remit, that they take place in the here and now but have a science fiction situation. Otherwise, they should have a futuristic or outer-space setting. Within the general production notes of the series, other factors in producing science fiction stories for television were to be considered more important besides the actual story value. Their literary interest, a note claimed, was practically the last concern.[29]

For example, Television Enterprises, the commercial arm of the BBC, was consulted about the potential for overseas sales, including the number of episodes required for a sale and the preferred running time. Twelve episodes for the first series were produced, running for sixty minutes. The second series was thirteen episodes running for fifty minutes, which was considered to be much easier to sell overseas. Catherine Johnson explains that twenty-five- and fifty-minute episodes were not only a key requirement for the domestic schedule, but also the only acceptable lengths for export in the 1960s.[30] Another problem for adapting literary material of variable length was the requirement on television for episodes of a standard length to increase the chances of foreign sales. As the BBC noted, "It is obviously better to record 51 minutes that can be cut, rather than 47 minutes which we cannot make longer." This form met the requirements of the television economy for a regular and extensive supply of programs that could be marshaled together to form a typical weekly schedule but also offer a mix of change and continuity. The literary value of the story had to conform to such extra-textual requirements.

By May 1965, the approximate budget for each episode was set at £6,000. In August, the stories had been chosen, and the plays were ready to go into production. The series was to begin transmission on October 4, 1965. Three stories had been recorded by the end of September, each one completed within five days. There would be two to three days in the studio and two days editing. At the time, anything transmitted by the BBC within fourteen days of its recording would automatically go onto tape.

Unfortunately for the new series, the first episode of *Out of the Unknown*, "No Place Like Earth," was accused of lacking drama and having too much dialogue. *Out of the Unknown* was described in *Radio Times* as "A series of new Science Fiction plays." Each episode was publicized, including photographic stills and a synopsis, beginning with a picture of Terence Morgan as Bert Foster, and a promise from Shubik for more "strange but plausible wonders," in future episodes. Yet the initial episode was not liked by the critics, perhaps because of the raising of expectations that were too high and too difficult to fulfill. About "No Place Like Earth," the *Times* was to comment:

> BBC 2 chose to begin its "exciting new series of science fiction plays" (as the *Radio Times* describes them) with *No Place Like Earth*, by Mr. John Wyndham, a writer whose work has all the social urgency and pertinence which devotees first found in the novels of H. G. Wells. Its hero, stranded on Mars after the end of the world, at first rejects the dying, untechnological graces, of Martian life to join a totalitarian, empire-building slave state on Venus. Disillusioned, he returns to the gentle, uncreative tolerance he had not previously appreciated.
>
> Space travel is, of course, a more ticklish subject than the sort of thing Mr. Wyndham's novels imagine happening on earth, but *No Place Like Earth* makes its social point clearly enough. Unfortunately, it does so extremely slowly and with heavily sententious dialogue underlining what is perfectly clear without its assistance.
>
> There was little for actors to do except strike attitudes; Mr. Terence Morgan struck all the right ones with proper discretion. Mr. Peter Seddon, who designed a suitably imaginative Martian landscape, was given little opportunity to add other visual excitements. Martian architecture, the work of a dead race, seems to be influenced by ancient Egypt.[31]

The choice of John Wyndham for the opening episode of the new series reflects the difficulty of establishing clear generic borders in science fiction generally and the specific problems of adapting what has been science fiction for television.

Wyndham had been born with the portentous name of John Wyndham Parkes Lucas Beynon Harris in comfortable middle-class English surroundings. He had been publishing short stories and serials in American science-fiction pulp magazine since the 1930s. After the war, and especially since the publication of his first significant novel, *The Day of the Triffids* (1951), his books dealt with disaster and the destruction of civilization, only to be confronted with the problems of what replaces it. The literary strategies and ideas in his stories reveal an inheritance from the American pulp magazines to write a gripping story. At the same time, author Brian Stableford argues that Wyndham was "the most striking example of a writer who contrived to combine the traditions of British scientific romance and American science fiction."[32] But Wyndham was not immune from criticism. Brian Aldiss argues that Wyndham's novels "were totally devoid of ideas but read smoothly, and thus reached a maximum audience, who enjoyed cozy disasters. Either it was something to do with the collapse of the British Empire, or the back-to-nature movement, or a general feeling that industrialization had gone too far, or all three."[33]

Significantly, when his books were reissued as paperbacks by Penguin, Wyndham insisted that they appear without the science fiction label.

> He did write for science fiction magazines in Britain and America—particularly for Carnell—but he made every effort to prevent his books being marketed as category fiction, preferring instead to cultivate an independent reputation. This he managed to do, and gathered a considerable following among readers who otherwise steered clear of labeled science fiction.[34]

Wyndham's rejection of the science fiction label for works he clearly wished to be regarded as literary, and which transcended genre, echoed the earlier need felt by the *Radio Times* to list *Quatermass and the Pit* as a "thriller." However, if the attempt to categorize texts satisfies the need to group particular texts together in order to successfully market them to a large audience, then crucially, categorization on television also helps to differentiate individual texts.

Categorization is central to television, and production and consumption are regulated by repetition if modulated by acceptable difference.[35] This was most obvious in the production of the serial form on television, regardless of its genre. The television series was an ideal form for the rationalized mass production of television narratives, and by the mid- to late 1950s a different episode of a series would be scheduled for broadcast at the same time every week. Moreover, the series and the serial were specific forms of narration that relied on scheduling, although the serial implied some narrative progression, while the series usually did not. As we have already seen in earlier chapters, the general principle of seriality generated many segments from basic thematic material, avoiding a single narrative arc, and allowing for longevity. The anthology series did allow more differentiation than repetition because a new adaptation appeared every week, but each type of narrative form allowed the act of viewing drama to be made routine by scheduling. The development of scheduling was due to ideas about the centrality of the family, leading to an assumption that drama would be made for a mass audience settling down at the same time in front of the television.

However, a totalizing account of the "mass" audience avoids the more obvious changes in taste and fashion of individuals sitting at home. Philip Purser, the TV critic of the *News Chronicle*, recognized this seemingly contradictory attitude when he observed:

> The question that matters is whether proper discrimination and genuine creativity can survive amid such a vast supply and demand. Can Jacques Gillies [a thriller TV dramatist] outlast *The Avengers*? . . . The public otherwise seems to be extraordinarily catholic in its taste, and also innocent in that it has few preconceived likes and dislikes. . . . The ratings are not much more helpful. A good play on ITV will usually make the Top Ten, but then *Wagon Train*, a lousy Western, is there every week.[36]

The tension between an industrialized medium and textual authorship on television had led Nigel Kneale to become contemptuous of the half-hour self-contained series, in which he claimed nothing worthwhile could be accomplished.

> The series, however, he regards as a complete play of 180 minutes, and he writes it as such. It builds steadily, each emotional climax being bigger than the last, corresponding roughly to the act and scene-breaks of the stage play. Best of all, it gives room for "suggestion": planting ideas which will act as springboards for the personal imagination of the viewer.[37]

Yet even the success of *Quatermass and the Pit* has to be put into the context of viewing figures and the "ratings game," with the BBC estimating some eleven million viewers watching the last episode in January 1959.[38]

"NO PLACE LIKE EARTH"

Wyndham's short story "Dumb Martian," originally published in 1952, had already been chosen by Sydney Newman for his earlier ABC series *Out of This World* because Wyndham was well known and had enjoyed previous commercial success. Two of his best-known stories had been filmed, *The Midwich Cuckoos* (1957) by MGM in 1960, as *Village of the Damned*, and

forms of narrative on television. Helen Wheatley has discussed some of these experiments in style and innovative and creative approaches to the rendition of psychological, "subjective realism."[43] She quotes Leonard White, the series producer of the spin-off of *Armchair Theatre*, *Armchair Theatre Mystery* (1960; 1964–1965) about style:

> How risky it was to experiment with changing the style of conventional television drama production. While still harnessed in the live method of recording productions, we were pushing all the time for the fast-moving style of film cutting (editing) with many and shorter scenes in the format—especially in [*Armchair Mystery Theatre*]. All of which constantly challenged the dexterity of the cameramen (and cable bashers), the boom operators and the "switcher," and crew in the control room. . . . Within the constraints of live recording we were living dangerously.[44]

However, such a frenetic pace visible in *Armchair Theatre Mystery* on ITV was absent in "No Place Like Earth."

The ambiguity at the heart of Wyndham's writing, rather than a matter-of-fact invocation of science fact, is used to stimulate reflection by the audience after Foster has listened to a tub-thumping speech from one of the Venusian colonists about rebuilding civilization. The notion that the planet Venus is "hot" is a reminder of the British colonial experience in other hot locales. At the same time, the episode is not devoid of innovative technique. The representation of Foster's conflicted, subjective state, as he contemplates the choice offered by the two planets and his desire for Annika's daughter, Zaylo, is constructed using a montage of complex superimposition rather than cuts. The episode calls on the vision mixer or "switcher" to create an extended montage of superimposed images. At first the camera tracks toward Foster for a close-up, before there is a superimposition to Annika within his line of sight. Voice-over reveals his reservations about the offer to travel to Venus. This conveys not simply an interior monologue by Foster, but the ideas and thoughts of the writer: Wyndham's authorial voice as an expression of the English intellectual tradition. The montage is multilayered and becomes a stream of consciousness as its drifts to Zaylo and thoughts about "the mania of owning things," before embarking on a more robust meditation on what civilization offers. There is a complex soundtrack, consisting of the sound of lapping water and contemplative music, before a dissolve takes the audience to film of Max Berg's strikingly modernist Centenary Hall in Breslau,[45] accompanied by the strains of the "Ode to Joy" from Beethoven's Ninth Symphony. The choice of a monumental building and soaring, orchestral music is interesting too, so soon after the architectural triumphalism of Nazi Germany. A final dissolve returns the camera to Foster. This impressive short montage of dissolves and superimposition conveys sensations of extreme emotional states, and this is clearly how Foster feels. Within the sequence, the appeal of domesticity, which can be represented by women, is countered by the appeal of the masculine construction of the nondomestic world: the achievements of civilization and its triumphalism. Paradoxically, each set of values attracts Foster, and using a complex style, which can be described as televisual rather than filmic, "No Place Like Earth" is able to signify alternating emotional states such as desire and anxiety, and, ultimately, the exhilaration of civilization as the precursor of an ego-driven madness.

According to the BBC's Audience Research Report, a manager explained that "the idea was interesting but in its development did not provide a gripping play."[46] An advertising representative opined that "As entertainment it was mildly acceptable, on the level of Dr. Who, and I wondered with some surprise who had decided to present it to an adult audience, when younger viewers would probably have reveled in it, even with Daleks."[47] The report concluded, "Turning to the production, some of the sample were disappointed with the settings, various viewers maintaining that they were "shabby," "stagey," lacked variety and "did nothing to suggest the events took place anywhere but on earth."[48] Yet, as we have seen, the episode is a complex narrative about the self-conflicted appeals offered by civilization. It orchestrates a multitude of layered images that extends the original authorial voice, which embellishes the original theme of the short story by Wyndham. The montage also creates a more subliminal relationship between the main character and monumental triumphalism, which ends in catastrophic war.

Mark Ward explains that, "The show failed to retain the attention of many; general opinions found it to be slow, unexciting, unconventional, not enough action, a sense of urgency was lacking."[49] Such reactions reveal a desire from those both inside and outside television for a type of storytelling that permitted more action. However, Wyndham's writing is imbued with a very English sensibility, which the BBC adaptation continues to inflect, avoiding the uncertainties of catering to foreign or US public taste but preferring to maintain the tone of the original writing. A BBC memo discusses Wyndham's fascination of the uncharted.

> Wyndham says that Mars today is to us what the New World was to men in the 17th century. It is just within reach, but we really know little about it. In *No Place like Earth* we have two unknown worlds for the price of one. Venus too becomes the scene for Bert Foster's travels. How does Venus differ from Mars, both geographically and politically? Will the Venusians really be able to profit from the mistakes which led to Earth's final destruction, or will the same cycle of events recur?[50]

ship. The internal rhythms of Westcott's disturbed mental state are matched to the alien sounds and patterns of light from the instrument panels on the bridge. The sequence continues as a montage, the shots separated by dissolves as they oscillate between Westcott and the ship. To heighten the growing confusion of his mind, images of the strange shapes and lights of the ship are superimposed to represent Westcott's disturbed psyche in a non-naturalistic way. The complex montage therefore subjectivizes a point of view around a technological rather than a human center. The images of the bridge of the ship are no longer anchored to a human POV, but create a strangely floating, omniscient point of view using a mobile camera to suggest high-tech artifice. Finally, the sequence rapidly cuts between Westcott and the fantastic shapes and lights inside the spaceship. Odd noises form a dissonance and end with Westcott crushing the ball into an oozing mess in his hand.

The entire sequence runs for about five minutes and demonstrates the ability of production at the BBC to extend far beyond a social realist or naturalistic paradigm. *Out of the Unknown* was created for BBC-2, a channel that had been mandated to extend the range and diversity of drama from BBC television, and stimulate the viewer's imagination through the development of character and theme contained in the original short stories. Moreover, the already established narrative sophistication of the original stories enabled *Out of the Unknown* to stimulate the imagination by its deployment of a range of styles, including montage, which unlike the social realism visible in *Armchair Theatre* and elsewhere, was able to draw on further heterogeneous possibilities.

The three most difficult sequences to create in "The Counterfeit Man" were to be the persecution montage; a pressure chamber sequence in which Westcott dissolves into a jelly in various stages; and the final sequence in which Doctor Crawford is counterfeited and plays a scene with himself. The second of the three montages was made up of seven shots, of which three had to be prefilmed, at a time when editing on video was extremely difficult without the aid of time codes and the ability to identify individual videotape frames. The difficulty of editing the sound in "The Counterfeit Man" suggests what the benefits of shooting a play entirely on film may have been. Without the facility to edit, the videotape recorder had been like its live predecessor not a serious contender to shooting on film. The breakthrough for video editing was the invention of the time code, the ability to identify every frame of videotape with a simple number that could be traced without difficulty and be used as a reference point for an edit. Time code was first used by the Electronic Engineering Company of America (EECO) and displayed elapsed time in hours, minutes, and seconds, and finally video frame numbers.[55]

NEW DRAMA AT THE BBC: *STRANGER IN THE FAMILY*

Meanwhile, the play that marked—if not the end of the studio play—the possibilities of using 16 mm film extensively outside the electronic studio was *Up the Junction* (1965), produced for the BBC and directed by Ken Loach. It had been enthusiastically welcomed by Troy Kennedy Martin, who explained why the "New Drama" had to embrace new technology and move away from the constraints and confines of the electronic studio:

> With conceptions as complex as Ken Loach is trying to realise, a director cannot cue in and out three sound tracks and the cameras and run in telecine with dead accuracy over the period of an hour or so. Such orchestration requires physical facilities—superhuman responses, as well as studio time—which he has not got. The only solution is the possibility of subsequent editing afforded by telerecording.[56]

The decision by Loach and Tony Garnett to shoot on 16 mm film had been a determination to represent a new type of realism in a drama about contemporary life in south London. Loach and the program's script editor had wanted to get out onto the streets, into the real world. The third episode of the *Out of the Unknown* series, "Stranger in the Family," was broadcast on October 18, 1965, a month before *Up the Junction* on BBC-1, but parallels exist, which suggest how identified stylistic qualities can be understood to be personal predilections, as well as changing production conditions at the BBC. The third episode is one of the few *Out of the Unknown* episodes not adapted from an existing literary work. Instead, it was written for the screen. This appears to be an admission that many preexisting short stories were often too condensed to be adapted as a sixty-minute teleplay.

David Campton was a prolific British dramatist who had written for several types of media, including stage, radio, and cinema. He had also worked on various anthology shows for television in the 1960s and had penned "Stranger in the Family." The central character of the episode, played by Richard O'Callaghan, is Charles, or "Boy" as he is called by his parents. Charles possesses powerful mental abilities and controls the actions of others by the power of his mind. He is also being tracked by a mysterious surveillance team who inhabit the flat next to his in a tower block. Charles falls in love with a young actress and her agent hopes to exploit his powers to make money from television commercials. The story may have been inspired by the short story *It's a Good Life* (1953) by Jerome Bixby, which became the basis of a 1961 episode of *The Twilight Zone*. The story is also thematically similar to John Wyndham's *The Midwich Cuckoos* (1957), which had been made into a film in 1960 as *Village of the Damned*. Each of these stories creates a series of parallels with other textual examples, but it is impossible to be certain if Campton chose to mine these earlier stories for his own fantasy.

Unlike the first two stories from the series, "Stranger in the Family" is set on present-day Earth. The generic verisimilitude of spaceships and aliens is absent. Instead, there is a clear interest in the here and now world that elsewhere at the BBC was being pioneered as the "New Drama." The opening of the episode is shot in the streets of London and develops a clear sense of place, a credible geographical location, as opposed to the illusion of space in the electronic studio. By creating such a place, it is possible for a personal narrative to unfold. We see Charles on a walking tour through London, beginning in the Science Museum, Kensington.[57] The scene cuts to outside: we see Charles continue to traverse the urban landscape and cross a Dickensian passageway to an area overlooking the River Thames. As he walks, we notice for the first time that he is being watched by a man. Charles's boyish enthusiasm for the city as a cavalcade of delights is menaced by his unknown stalker. The style of the opening is reminiscent of the location shooting of the French Nouvelle Vague, including its lyricism, its honesty, and the clarity of the streets of London. The world of film would have been a possible influence on a playwright as versatile as Campton, who was also known for producing a type of play labeled "comedy of menace," which was closely linked to the work of Harold Pinter.[58] The play continues with additional shots of Charles sitting down and overlooking the river, before cutting to a busy road. There the man who has been following Charles, wordlessly approaches him as Charles shouts for him to go away. The Pinteresque encounter leads to the man being hit accidentally by a truck, and Charles makes a dash for it. As he runs to the block of flats he lives in, another menacing figure follows him in the distance. Charles cannot escape the mysterious force that is keeping him under surveillance, and his grasp at freedom is thwarted. As he uses the walkways of the block of flats, the camera adopts his POV of the city beneath him. Charles walks into the flat to meet his parents and, after six minutes running time, the scene shot on film ends and the play continues with what was recorded inside the electronic studio.

"Stranger in the Family" reveals several things happening within television drama. Rival discourses between television as an electronic medium and program makers deploying 16 mm film was leading to a diversity of forms being used in British television drama serving a public service and dramatic interest. At the same time, the technology of videotape was catching up with the flexibility that film could afford. For example, videotape editing offered obvious advantages in the elimination of mistakes during the broadcast of a program. According to one television pioneer in this regard, "Recording before transmission allowed studio faults to be corrected and all material to be post-edited in the cutting-room."[59] In the course of the 1960s, the transmission of live fictional material practically ceased. It became normal procedure to record onto video intermittently, rather than in one continuous take, and to reorder material after the event. Electronic studio television was moving closer to the ability of film to record and resequence material, and there was a parallel growth in the actual use of film within television. Denis Forman describes these two developments—videotape editing and the use of the film camera—as happening simultaneously.[60] The question that was faced in the mid-1960s was not whether video was more efficient than film, but how far VTR should be used to "open out" a TV play by adopting some of film's visual vocabulary.

Twenty years later, in his MacTaggart lecture at the 1986 Edinburgh Television Festival, Troy Kennedy Martin felt that his attack on the dramatic style of naturalism remained valid, and he cited the role of technological developments over this period as being crucial to naturalism's dominance.

> One of the perennial problems we have had to face every time we have been confronted with a situation which calls for *imaginative change* is that we are let off the hook by new technological developments which allow the old way of doing things just a little more life. We started in the drama studio with black-and-white, then went onto colour, then out into the streets with mobile tape, then film, then Super-16, then faster film, then 35mm. . . . At each stage, when the process should have been thrown back onto the virtuosity of its creators, the new development has allowed the Establishment to keep pumping out the same old naturalistic tune.[61]

In other words, Kennedy Martin was asking whether there would be a phenomenological use of audiovisual space rather than an aesthetic use of moving images. Initially, Kennedy Martin believed a "new drama" for TV could be achieved in the television studio rather than on film because he believed that naturalism as a dramatic form had visually evolved from Hollywood film techniques.[62]

This essentialism of television and film form is complicated by the fact that, by the 1970s, video was an electronic medium that had been constantly developed for an electronic system in the same way that film had been originally designed for big screen front projection. However, conversely, many electronic cameras were fitted with film lenses in order to make use of scrims, diffusers, nets, and the short depth of field that had been traditionally used on film cameras.[63] Crucially, instead of an electronic method of recording and a celluloid method of filming there was a convergence between an electronic and film medium. Unlike film, the sharpness of the image from a video camera could suggest liveness, but increasingly video was used like film with the aid of film lenses to become more of a sophisticated or complex aesthetic medium that could be compared to the use of film at the cinema.

A major factor in the success of new forms of drama in *Out of the Unknown* was the widening coverage of BBC-2 to more of Britain. BBC-2, first introduced in London and the Southeast of England in April 1964, had been extended to the Birmingham area in the autumn of 1964 and then to many parts of the country during 1965. By 1967 it covered around two-thirds of the population. A choice of programs on BBC-1 and BBC-2 continued to be planned on a complementary basis. One of the benefits of such planning was the common time junctions at which viewers could switch from the end of one BBC program to the start of another on the other channel. Television imposes logic and organization on flow through scheduling, but complementary scheduling strove not to constrain the determinations of the plays in *Out of the Unknown*, whose appeal continued to be its eclecticism.

The ability to supersede the parameters that had hitherto made up science fiction on television was praised as the reason for watching the show. In a letter to Shubik from *Vector*, the official organ of the British Science Fiction Association, in the winter of 1965, Martin Pitt explained:

> I think that the series was a major turning point in the history of dramatised science fiction. The old idea of pop science fiction with element X, ray-guns and silver space suits which everyone wears all the time, is still current: viz "Lost in Space," but you have, thank heaven, broken away. . . .
>
> The special effects such as the death of the *Counterfeit Man* are what everyone is talking about.[64]

After the success of episodes such as "The Counterfeit Man" and "Stranger in the Family," the series continued to gain in popularity until it became the most successful drama after the US import *The Virginian*.[65] The average audience reaction index (RI) for television plays and serial episodes was 61, and "The Counterfeit Man" had received an RI of 63, and "Stranger in the Family," an RI of 68. After its initial run, a decision was made to repeat the first series on BBC-2 before broadcasting the second series. "No Place Like Earth," which had received an RI of 53 on its initial broadcast, was able to achieve a rating of 63. The play's philosophical theme had more of an appeal on its repeat and indicates how television can build an audience using the communal experience of watching a particular show. The appeal of television that is worth talking about and its association with informed viewers who possess sufficient cultural capital to understand its complex themes helps to explain the increased popularity of the repeat of "No Place Like Earth."

BACK TO THE FUTURE: "THE MACHINE STOPS"

By 1966, the space race had progressed to the missions of Project Gemini, which featured complex rendezvous and docking operations, and spacewalks. The success of such missions meant that the American manned space program was ready to begin preparing the Apollo series of missions to go to the moon. Therefore, it seems odd that the first episode of the second series of *Out of the Unknown* was a reversion to a historical form of science fiction. "The Machine Stops" was a short story originally by E. M. Forster who had written it before World War I "as a reaction to one of the heavens of H. G. Wells." The story is an Edwardian view of the future, and Forster saw—sooner than many writers of his time—the dystopia of a mechanized society. Kenneth Cavander, a classical scholar who at Oxford had translated Euripides' *Hippolytus*, adapted "The Machine Stops" to television. The play was directed by Philip Saville, a prolific and stylistically innovative director, who was known for his rapid and intricate camera movements. His work was also noted for its interest in psychological states and subjective viewpoints.

The story is about Vashti (Yvonne Mitchell) and her son, Kuno (Michael Gothard). Vashti is hairless, mechanical, and virtually sexless and permits the Machine to control her life and the lives of others in a mechanized future. One day, Vashti is contacted by Kuno, who asks her to visit the other side of the Earth with him. Unlike Vashti, Kuno is anxious not to conform to the strictures of the Machine, which he starts to reject. For him the Machine, originally designed to serve, has become the master and is venerated as a god. It provides food and drink, sex and bed, music and verse, education and instant television communication all over the world. To the Machine, Kuno suffers "from the demerits of muscularity." His physicality is at odds with the sedentary Vashti. Kuno's application to become a father is rejected because his physical strength is considered by the Machine to be a disadvantage in a world of enervated humanity. Kuno decides to visit the surface of the Earth and he sees a girl, but before he can talk to her, robotic arms in the shape of giant worms reclaim him and return him to the vents he used to climb to the surface. The script is faithful to the original story. The only major departure from it is that Kuno's visit to the outside world is told in flashback in the short story after Vashti arrives in his cell. Meanwhile, Vashti has chosen to fly to the other side of the Earth, but bemoans her terrible journey and the fact that she was almost touched by a beam of sunlight. Two years later, Vashti is in her cell, and has begun to notice small defects that the mending apparatus of the Machine has failed to rectify. She is contacted by her son, who has been held in his cell for two years. He has had a thought which he believes will frighten his mother: "The Machine stops." The dissolution of the Machine-operated society continues until the Machine falls silent. In the ensuing panic Kuno is struck by a train and, in his final moments, mother and son realize the threat to human relationships from mechanization.

Saville uses montage when Kuno suggests to Vashti the possibility of the Machine stopping. A number of cutaways are edited together in rapid succession with intercuts of instrument panels. Duplicate sets are used throughout the production to allow for multiple camera angles, which was a technique that Saville specialized in. The hexagonal shapes of the set represent the description of the cell in the original story. The pulsing lights and thin membranes suggest that the machine is alive. Stock sound effects are featured that had previously been used in *Doctor Who*, including the operating noise of the doors that had been heard in early Dalek adventures. One idea from the book, which was in the script, but did not materialize on screen, was to end on a shot from the ceiling above Vashti and Kuno. The ceiling would disintegrate to reveal the sky overhead and form a dramatic finale. However, due to costs, the production ends on a two-shot of the main characters with a bright key light on them. Smoke effects and strobe lights are also deployed for the chaotic finale.

The reversion to a much earlier type of science fiction seems at odds with the show's established policy of adapting stories with a contemporary setting. Originally, Saville had expressed the wish to direct the play in an Edwardian style or using the Edwardian idea of what the future would look like. Consequently, the décor and settings emphasize the use of monorails, and the hexagonal shapes of the scenery are meant to be the remembrance of a Wellsian futurescape. Shubik was to recall that "The Machine Stops" was the most difficult television production she had overseen.[66] When it had been recorded, an extra hour of recording time had been granted because of the complexity of the production. It was a pessimistic Edwardian view of the future, an attack on the tyranny of the "machine age," which Forster believed Wells had championed and he had critiqued in his short story. The show's interest in Wells and his prophetic view of the future had been due to the centenary of Wells's birth in 1866:

> Period science fiction has been much in mind lately because of the Wells centenary. Half the appeal, of course, is the way in which yesterday's soaring visions of the future are nevertheless rooted in the materials and technology and preoccupations of their age. I thought the designer, Norman James, preserved a good deal of this in *The Machine Stops*, from E. M. Forster's 1909 story, but might have preserved even more . . . the spaciousness of the "aero-ship" with a handful of seats dotted around a vast deck, truly Edwardian.[67]

Forster's attack on Wells in "The Machine Stops" can be interpreted not simply as a piece of anti-Wells propaganda, but as an attack on many tenets of scientific romance: an opposition to its assertion of the triumph of rational thought against the irrationality of the preindustrial age. For Forster, industrialization threatened to lead to a narrowing of the individual, whose own unique value would become diminished in the Age of the Machine. The story is therefore able to function as a critique of the centrality of categorization and the standard form of generic boundaries. Such boundaries can only exist if we accept a clear determinism—machine-like—that ignores the possibility of individual contributions. Instead, it suggests that in similar ways the primary sense of creation in television clings to the word as the direct expression of the individual author.

The hungry maw of TV for scripts had led to the adaptation of a story that somewhat paradoxically attempted not to be television. Rather, the *Daily Telegraph* was to acclaim the fact that "Enough Forster was left to make the dialogue unusually distinguished."[68] However, as we have already seen, television is an industrialized medium and a series such as *Out of the Unknown* cannot ignore that it deploys an ideal form for the rationalized mass production of television narratives. The use of a form such as the television series limits the range of subjects the contributing writer can select from. On the other hand, categorization on television helps to differentiate individual texts such as Forster's.

CONCLUSION

This chapter has explored how some of the problems of adaptation are evident in the tension between television as an industrializing medium and the concept of a single, creative, originating source for each of the short stories in *Out of the Unknown*. Shubik's original idea had been that *Out of the Unknown* would be basically a series of single plays. Her role as producer would be to ask for stories that could conform to existing themes, character types, and narrative models. The concern would always be with science, technology, and material things. However, the ideas would not simply use stereotyped characterizations. Rather Shubik's supervening role would also be a collaboration between the producer and the director, as well as further collaborations between the adaptor of the short story, and a host of other staff, including script editors, actors, and technical crew. The range and usage of creativity on television can be extrapolated to suggest further categories of "distinction" or "quality." Within this model of viewing, the television audience was likely to be differentiating between separate items on BBC-1 and BBC-2. Within this scenario, the bounds of distinction continue to be important and ideas about the writer encourage clear use of the deployment of visual codes that mark "The Machine Stops" as more than "science fiction." The show's engagement with its audience remains complex as an opposition to the ideological basis of genre. In the hands of Philip Saville, the director, the tale of "The Machine Stops" subverts the notion that scientific fiction will "discover the future." In other words, "The Machine Stops" should be understood as a thought experiment about how technological developments can be detrimental to the future of social behavior. This interest is not extemporaneous, an Edwardian curiosity. The episode is about television itself. The creation of the ultimate personal space of the "cell" in "The

Machine Stops" acts as a cipher for the domestic space occupied by television.

NOTES

1. Cyclops, "A for Andromeda," *Television Mail*, October 13, 1961, 13.
2. Shubik, "Producers and Directors," 87.
3. Ibid., 88.
4. Quoted in Ward, *Out of the Unknown*, 14.
5. Alfred Bester, letter to Irene Shubik, February 16, 1965, BBC WAC T5/1,755/1.
6. Ward, *Out of the Unknown*, 25.
7. Memo from Shubik, BBC WAC T5/1,755/1.
8. Tony Gruner, *Kinematography Weekly*, August 9, 1962, 16.
9. "Imposter," *Out of This World*, originally transmitted July 21, 1962.
10. Gruner, *Kinematography Weekly*, 16.
11. This includes the first episode, "Dumb Martian," which was made for *Out of This World*, but broadcast as part of *Armchair Theatre*.
12. Honri, "Make Sure We Sell," 11.
13. Ibid, 11.
14. See Kotcheff, "The Making of Life at the Top."
15. Wheldon, "BBC Executives," 223.
16. Hill, "Creative in Its Own Right," 18.
17. Ibid., 17 and 24.
18. Ibid.
19. The Langham's final production (it produced three altogether) was broadcast in July 1960.
20. Kneale, "Not Quite so Intimate," 187.
21. Sexton, "Origin of Gritty Realism."
22. Todorov, *The Fantastic*, 28.
23. Schatz, "The Structural Influence"
24. Feuer, "Genre Study and Television."
25. Johnson, *Telefantasy*.
26. The phrase was used by Prime Minister Harold Wilson during his address at the Labour Party conference in October 1963. See Sandbrook, *Never Had It So* Good, 691.
27. Irene Shubik, untitled document, BBC WAC T5/1,755/1.
28. SWFA Bulletin, January 1966, 10.
29. BBC WAC, *Out of the Unknown*—general, T5/1,755/1.
30. Johnson, *Telefantasy*, 44.
31. "Space Travel at a Walking Pace," *Times*, October 5, 1965.
32. Stableford, *Scientific Romance*, 326.
33. Aldiss, *Trillion Year Spree*, 315.
34. Stableford, *Scientific Romance*, 326.
35. Caughie, "Adorno's Reproach." Caughie discusses difference not to mean simply variation, but the negation of the rationality of capitalist production and bourgeois ideology as first proposed by Adorno and Horkheimer.
36. Purser, "Landscape of TV Drama," 12.
37. "Intelligent Viewers Must Be Captured," *Television Mail*, November 13, 1959, 16.
38. Audience Research Report for *Quatermass and the Pit*, episode 1, "The Halfmen," BBC WAC T5/2,302/1.
39. *The Midwich Cuckoos* was originally published in 1957 and *Day of the Triffids* in 1951. The latter was also dramatized for radio by the BBC in 1957.
40. Ward, *Out of the Unknown*, 30.
41. Ibid.
42. Vincent Price had been approached to introduce the episodes.

43. Wheatley, "And Now for Your Sunday Night Experimental Drama . . . : Experimentation and *Armchair Theatre*," 40.
44. Ibid., 41–42.
45. The author cannot be certain of this identification, but it looks like it.
46. Audience Research Report, *Out of the Unknown*, episode 1: "No Place Like Earth," BBC WAC T5/1,725/1.
47. Ibid.
48. Ibid.
49. Ward, *Out of the Unknown*, 42.
50. BBC WAC T5/1,725/1.
51. Originally broadcast October 11, 1965.
52. *Radio Times*, July 2–8, 1966, 43.
53. *Daily Sketch*, October 12, 1965.
54. Kennedy Martin, "Nats Go Home," 24–25.
55. Anderson, *Video Editing and Post-Production*, 6.
56. Kennedy Martin, "*Up the Junction* and After," 140.
57. This is impossible to be absolutely certain about, but the sight of the exhibits and nature of the program (science fiction) do strongly indicate this.
58. Campton had performed in the second production of the *Birthday Party* with Alan Ayckbourn and Pinter.
59. Barry, *From the Palace to the Grove*, 174.
60. Forman, "Television in 1959," 60.
61. Kennedy Martin, "Sharpening the Edge of TV Drama," 12.
62. Kennedy Martin, "Nats Go Home."
63. Morton Dunn, "Commercials," *Broadcast*, April 26, 1976, 14–15.
64. Martin J. Pitt to Irene Shubik, December 17, 1965, BBC WAC T5/1,755/1.
65. Ward, *Out of the Unknown*, 32
66. Shubik, *Play for Today*.
67. *Sunday Telegraph*, October 9, 1960.
68. *Daily Telegraph*, October 7, 1966.

Chapter Four

Alternate Histories

Animation in 20,000 Leagues *and* Superman

We have already seen how animation played an important role in the adaptation of science fiction to television within the *Disneyland* show, where it acted as a bridge between the science and fiction in Disney's foray into television. Animation's ability to present historical and factual material in a clear and precise manner was put to use while its association with entertainment made the subject matter accessible and offered a place for the more fantastical elements of Disney's exploration of space.

In the 1960s, animation would continue to prove to be an important area for the development of science fiction adaptation for television. Despite Disney's strong association with animation, at this time the company was pulling back from its investment in and reliance upon animation production, as demonstrated by the *Disneyland* show and the theme park it supported, as well as the increased production of live-action films rather than animation, notably *20,000 Leagues Under the Sea*, discussed in detail in the second half of this chapter.

Instead a new group of companies, including Filmation, Hanna-Barbera, and Rankin/Bass, emerged within the television industry to exploit animation in a very different way. For these companies science fiction would be an important area in which perceived medium specificities could be challenged and negotiated. These companies adapted science fiction shows from diverse sources, including live action film and television, radio, literature, and comic books, thus addressing the translation of material between these media. Furthermore, in doing so they also questioned the boundaries and definitions of animation, science fiction, and television, changing the complex relationship between these three forms.

TELEVISION AND ANIMATION IN THE 1950S AND 1960S

The thirty-year period before the rise of television as a mass medium was one of relative stability for animation, described as its "golden age" by Michael Barrier among others.[1] The field was led by Disney, who, in the wake of the success of Mickey Mouse's debut in *Steamboat Willie* in 1928, had grown exponentially, both in the production of theatrical short films starring Mickey, Donald Duck, and Goofy as well as expanding into feature film production, starting with *Snow White and the Seven Dwarfs* (1937). The major Hollywood studios followed suit, initially contracting with outside animation studios to produce films for their distribution and then establishing in-house units by taking over their partner; MGM, Warner Bros., and Paramount all followed this route. Walter Lantz (whose cartoons were distributed by Universal) and Paul Terry's Terrytoons studio (whose films were distributed by 20th Century Fox) remained independent operations throughout the period, but closely associated with the studio that distributed their work. Each of these companies established a regular production of theatrically released short cartoons, typically seven minutes in length, providing supporting material for the dominant narrative feature and the overall cinema program.

This "golden age" of animation came into crisis in the 1950s due to a complex set of factors, resulting in the closure or radical restructuring of all of these animation studios. The Supreme Court's 1948 ruling in *United States v. Paramount Pictures*, which forced the Hollywood studios to divest themselves of theatre chains and disrupted the vertical integration that had underpinned their financial success, is commonly cited as the key turning point.[2] The Supreme Court decision undoubtedly played a role in the decline of the production of animated theatrical shorts, but it cannot be considered the only factor. Production costs for animation had been rapidly rising since the late 1920s due to the increasing complexity of the production process with the introduction of sound, color, visual intricacy of the image, and increasingly sophisticated character animation. These were all labor-intensive activities that required skilled staff at a time when wages and working conditions were being closely scrutinized, as exemplified in the Disney strike of 1941. The rise of television provided a further factor in shifting the marketplace for animation, coinciding with and contributing to changes in audience patterns.

Examining in detail these structural changes is outside the scope of this book, but the results of them underpinned the role animation would play in the adaptation of science fiction for television. These changes destabilized the definition of animation as primarily a short-form theatrical experience for a general audience, including adults, and based around "familiar barnyard animal characters"[3] like Mickey, Donald, and Goofy; Bugs, Daffy, and Porky; and Tom and Jerry. Of particular importance here, the period of studio production had not seen science fiction as an important genre for animation. There were some instances of science fiction creeping into the Hollywood studio theatrical cartoons: Warner Bros. cartoons featured Marvin the Martian, most notably in the science fiction pastiche *Duck Dodgers in the 24½th Century* from 1953, the same year Paramount/Famous Studios released *Popeye, the Ace of Space*. The gadget-obsessed Wile E. Coyote, who frequently uses space-age technology purchased from the ACME company to pursue Road Runner, might also be seen as a trace of the increasing interest in science fact and science fiction in the 1950s discussed throughout this book. Yet these are isolated examples from the vast catalog of theatrical cartoons produced by the Hollywood studios, which more frequently preferred to look to the past for inspiration, whether the recent past of vaudeville entertainment or more distant historical periods.

A key exception to this was Terrytoons' Mighty Mouse character, who debuted in 1942 in *The Mouse of Tomorrow* (billed there and for his first two years as Super Mouse).[4] Mighty Mouse was a pastiche of Superman, who was enjoying his first fame following his debut comic book publication in 1938 and the equally popular syndicated radio show, which began in 1940.[5] While the Mighty Mouse character was sufficiently commercially successful to continue appearing in regularly released theatrical shorts for over ten years, he was a minor cartoon character compared to the stars from MGM, Warner Bros., and Disney previously mentioned, and consequently Mighty Mouse has been given only passing attention in histories of cinematic animation. However, as a case study of the adaptation of animation for television, Mighty Mouse provides an example of the wider patterns seen in early television's use of animation and the increasing role of science fiction within it.

The earliest appearances of animation on network television in the United States were adaptations of existing theatrical cartoons from the major film studios of the "golden age." Paul Terry's Terrytoons were one of the earliest examples of this exploitation of existing film libraries by television. The established Hollywood studios were wary of television, which they saw as a competitor to their business, and were reluctant to sell their vast libraries for television broadcast. 20th Century Fox, unlike MGM, Warner Bros., and Paramount, did not own the cartoons they distributed, instead ownership was retained by the independent production company run by Paul Terry. Terry saw television as an opportunity rather than a threat and in 1953 sold 112 old cartoons to CBS for $140,000.[6] At this stage these were all old black-and-white nitrate cartoons from the 1930s and 1940s, such as those featuring the Farmer Al Falfa character, rather than the latest color theatrical shorts, and the Terrytoon name was not emphasized; rather, the show was labeled *Barker Bill's Cartoon Show*. This show pioneered the reuse of theatrical cartoons for television by adapting them for a longer time slot in which several cartoons were joined together with newly created linking sequences, in this case

of this study, this also provided a space for science fiction animation on television.

As suggested earlier, science fiction had not been a dominant source of material for theatrical cartoons. While sufficiently commercially successful to be kept in production as theatrical releases from 1942 until his transfer to television, Mighty Mouse was a minor character in comparison to the MGM, Disney, and Warner Bros. stars, and equally received limited critical acclaim: Terrytoons was only nominated for the Short Subject (cartoon) Academy Award ("Oscar") four times and never won in this category, which was dominated by the aforementioned other Hollywood studios. Undoubtedly this low esteem contributed to Paul Terry and his Terrytoons being early adopters of television adaptation: while the major animation units had much to lose by being associated with the new medium, Terry saw limited risk and considerable gain in selling his library to CBS. In adaptation to television Mighty Mouse became both a critical and commercial success, not only leading to the other Hollywood studios following suit and selling their back catalogs to television networks and for syndication, but also demonstrating a market for science fiction–influenced animation that would only grow in the cultural and social context of the 1950s and 1960s discussed throughout this book.

As Jason Mittel describes, the proven success of adapted theatrical cartoons on television prompted the production of shows made expressly for television. This shift was motivated by several factors. The finite archives of the Hollywood studios in the face of television's multichannel morning-to-night schedules meant reruns rapidly became a staple of the schedule by 1959 and new material was at a premium. As *Variety* reported, "Rerunitis has set in quite heavily."[24] The growing importance of color for television broadcasts further limited the value of cartoons produced in black and white, which constituted a significant proportion of prewar theatrical material. Finally the closure of the major Hollywood animation studios meant there was a pool of talented animation staff looking for employment.

The result was the growth of animation studios dedicated to producing low-cost original animation for television. In the "golden age" theatrical cartoons had budgets in the region of $65,000 for a seven-minute short.[25] Television networks were not willing to finance animation at this level, not only because they needed to fill much longer program slots, but they had also become accustomed to the low-cost reuse of older theatrical cartoons. Furthermore, as Jason Mittel shows, the networks saw the newly defined exclusive juvenile demographic as undiscerning and tolerant of low production values.[26] While Jay Ward had been producing *Crusader Rabbit* expressly for television as early as 1949, it was the pioneering work by Hanna-Barbera that introduced limited animation to meet this demand on a wide scale.[27] This involved adopting techniques to reduce the labor-intensive animation process and associated costs. Hanna-Barbera was founded by William Hanna and Joe Barbera in 1957 following the closure of the MGM animation unit.[28] As directors of the long-running *Tom and Jerry* series, Hanna and Barbera had achieved great commercial and critical success at MGM and were responsible for some of the most acclaimed "full" animation of the theatrical period, dominating the Academy Awards in the 1940s with their shorts. Yet the duo proved equally adept at changing the animation production workflow to suit the lower budgets of television, relying upon techniques such as reuse of animation cycles, simplified movement, and use of the audio track for dialogue as well as sound effects and music, to compensate for simplified visuals. Starting with the successful *The Ruff and Reddy Show* (1957–1960) Hanna-Barbera used this approach to produce a string of popular cartoon series. While they achieved some success with prime-time shows, especially *The Flintstones* (1960–1966), the Saturday morning window remained the most reliable outlet for this form of animation.

Other animation studios focused on the production of low-cost television shows followed Hanna-Barbera in the late 1950s and 1960s, including Filmation and Rankin/Bass (both discussed in more detail below) and others such as Cambria Productions, Soundac, Beverly Hills Productions, Larry Harmon Productions, Rembrandt Films, Snowball Inc. (run by famed former Warner Bros. director Bob Clampett), and Crescent Studios. The result, as *Variety* described it, was that it was "raining cats & dogs as cartoons tailored for TV go into high gear," with the cartoon business thought to be worth in excess of $35 million by 1961, including growing production of animated commercials.[29]

This growth in production provided a space for science fiction to become an important part of animation television programs in the 1960s, fulfilling the promise offered by the success of Mighty Mouse. Shows that were either fully committed to the science fiction genre or featured it prominently in some plotlines included *Colonel Bleep* (1957), *Spunky & Tadpole* (1958), *The Space Explorers* (1958) and its sequel *New Adventures of the Space Explorers* (1959), *The Jetsons* (1962), *Space Angel* (1962), *Rod Rocket* (1963), *Jonny Quest* (1964), *Underdog* (1964), *Atom Ant* (1965), *Astronut* (1965), *Roger Ramjet* (1965), and *Space Ghost* (1966). Each of these shows adapted science fiction tropes from live-action television, radio, theatrical film, comic books, and literature, and in doing so negotiated them with perceived medium-specific qualities of television animation. Despite the clear growth in the production of science fiction–themed television animation in the years leading up to 1966, it was the arrival of Filmation's *The New Adventures of Superman* (1966) that marked the culmination of the trends described here: the move toward animation specifically and exclusively intended for a juvenile audience and scheduled in the Saturday morning time slot, along with the importance of science fiction as a genre for negotiating the medium-specific qualities of television animation.

DEFINING ANIMATED TELEVISION: *THE NEW ADVENTURES OF SUPERMAN*

As we will see, *The New Adventures of Superman* was both a beneficiary and cornerstone of the new paradigm of animation on television, but it was not the first adaptation of the comic book superhero, or even the first televised or animated version of Superman audiences had seen. Jerry Siegel and Joe Shuster had been developing the Superman character from 1933 when Siegel published a story "The Reign of the Superman" in a homemade magazine, which they titled simply *Science Fiction.*[30] After several years of attempts to get their character published, it was *Action Comics #1*, dated June 1938, that introduced Superman to the general public, with almost immediate success.[31] Superman was already the product of medium hybridity, Siegel and Shuster's influences ranging from the literature of Edgar Rice Burroughs (the Tarzan series of novels) and Phillip Wylie (the character Hugo Danner in *Gladiator* [1930]) to the films of Douglas Fairbanks Sr., science fiction magazines such as *Amazing Stories*, and other comic books and strips.[32]

Superman's ability to accommodate adaptation was demonstrated very quickly, as a daily newspaper strip was launched in 1939. The conventions of a four-panel black-and-white daily strip demanded a different narrative and graphic approach from a thirteen-page color comic book, yet Superman repeated his comic book success in the newspaper strip, quickly being syndicated to newspapers throughout America.[33]

The following year Superman was adapted to another medium, radio, which by 1940 had become a hugely popular mass entertainment form. Despite the obvious commercial appeal in such a move, the adaptation between two seemingly diverse media raises one of the central concerns of this book, regarding medium specificity. As suggested in chapter 1, radio was structured as a medium of broadcast or transmission, with audio transcription or recording technologies neglected in favor of the immediacy of live transmission relayed. In contrast comic books were seen as primarily a graphic form of communication dependent upon reproduction. How could such seemingly distinctly defined media both host the same character without departing from the derivation of his success? Yet *The Adventures of Superman* was a further success for National Comics (later to be renamed DC Comics) and, as such, indicates the way media could accommodate a wider range of qualities than narrow definitions allow. Key to this success lay in the flexibility, or adaptability, of Superman's dual identity.

Superman is a character of halves: his superhuman abilities resulting in spectacular storylines of rescue, while his disguise as mild-mannered reporter Clark Kent offers human interest, particularly in his interactions with Lois Lane. In turn these two sides may be seen to reflect upon fundamental tensions inherent in science fiction more generally: fact and fiction, narrative and spectacle, nature and culture. Radio offered a medium able to encapsulate both sides in equal measure. Radio placed no limitation on the locales and scale of adventure, as episode titles such as "Mystery in Arabia," "The Mayan Treasure," and "Airplane Disasters at Bridger Field" indicate.[34] The serial format with cliff-hanger endings was a staple of radio broadcasts, equally continuing plotlines were effective in the daily newspaper strip, so their adoption for *The Adventures of Superman* radio show was straightforward. Yet radio was also associated with immediacy thanks to its transmission of live voices into listeners' homes, well suited to exchanges between Clark Kent and Lois Lane. Central to the show's balancing of these elements was the capacity to enact the transformation of Clark Kent into Superman, thanks to voice actor Bud Collyer's celebrated ability to change the tone of his voice, audibly enacting the change as he said the catchphrase "this looks like a job *for Superman*."[35] This catchphrase, along with others, also gave Superman an identity that substituted for the loss of the iconic visual imagery that radio could not, of course, replicate directly. Catchphrases such as "Up, up, and away" served as a brand identity equivalent to the comic book's instantly recognizable red, yellow, and blue outfit, as well as describing actions that were tricky to depict audibly. It should also be noted that the radio show introduced Superman's weakness to kryptonite as part of his persona, a key part of the character's makeup that was only introduced to the comic book much later.[36] This innovation is a further reflection of the way the character was adapted for radio's perceived specificity, introducing a flaw that created a more accessible human character that would suit radio's intimacy. It was also a further indication that Superman, even at this stage in 1940, was a transmedial franchise in which the character straddled multiple media with none being considered the master source for adaptation, a pattern that would continue through further adaptations.

In 1940 Paramount and Superman Inc. signed an agreement for a series of theatrical animated cartoons to be produced by the Fleischer Studio in Miami, although the first release would only appear in 1941. The adaptation was clearly produced in the context not only of the comic books but also the daily comic strip, the radio show, and the expanding range of merchandising available, all of which were explicitly mentioned in announcements of the deal, as well as Paramount's advertising.[37] The immediate influence of the hugely popular radio show is apparent in the decision to use the same voice cast, including Bud Collyer as Superman, but the cartoons also needed to take into account the other adaptations as well as the qualities of animation as they were defined at that time. The Fleischer adaptation took key elements from these multiple sources, again reflecting the understanding of medium specificity, in this case of film and animation in particular.

Led by the Disney studio in the 1930s, animation had become seen as an art form of the expression of character through movement, best exemplified by the seven dwarfs in *Snow White and the Seven Dwarfs* (1937), each of whose movements expressed their inner personality. Spectacle was certainly a part of its expression but always to support the character-driven narrative, as seen in the use of the multiplane camera to provide a deep space structured around the character's performance.[38] The capacity of this form of animation to express complex thinking characters with personalities and emotions was clearly well suited to the depiction of Clark Kent, yet it also posed several challenges.[39] First, the character animation tradition, while concerned with psychologically motivated characters, had remained dominated by anthropomorphized animals or greatly caricatured humans and therefore was not immediately suited to the more realistic models used in the comic books. The use of rotoscoping (the tracing of live action footage onto cels rather than relying purely on the animator's drawing ability) could assist with this and was consequently used extensively for these animated cartoons, yet this contributed to the second challenge of animating Superman: cost. Achieving a high standard of character animation was labor intensive and therefore costly, and as a result the Fleischer *Superman* shorts had a reported budget of $100,000, a high sum even for the theatrical cartoons of the period.[40] The final challenge in adapting Superman into animation was the demand of visual spectacle associated with the superhuman elements of his character. With the radio show, comic strips, and books all in wide circulation, the attraction of an animated film adaptation lay in that form's other capacities: the translation of the bold, colored graphic design of comic books into movement and the capacity for spectacle that the large-scale cinema screen offered.

For the Fleischers these challenges undoubtedly contributed to delays in production of the cartoons: originally advertised as available by Christmas 1940, the films did not begin to appear until mid-1941.[41] On release, however, contemporary reviews found that they were successful in balancing the competing requirements between previous Superman adaptations and the specificity of animation, and between the competing demands of Clark Kent and Superman: the *Variety* reviewer favorably commented on both aspects: "jammed with action and the same implausibilities found in the newspaper strip. Better work on the human characters by Fleischer's staff makes this an above-the-ordinary entry."[42] As with earlier adaptations the Fleischer's Superman introduced elements that were then integrated into preexisting adaptations as well as future ones, including the phrase "faster than a speeding bullet," and the second episode "The Mechanical Monsters" was featured in a later Superman comic book.[43] The Fleischer's/Paramount cartoons were also the first adaptation to show Superman in flight, their meticulous character animation lending his movement a plausibility that would be difficult for later adaptations to follow.

A live–action film serial produced by Columbia followed the animated Superman cartoons as the next adaptation of the franchise, the first story arc released in 1948, and the second in 1950.[44] As with other adaptations, this version was working from multiple sources ("based on the comic strip and radio program" as one trade paper described it) as well as taking into account the expectations of its chosen form, the theatrical film serial.[45] As we have seen, seriality was already a part of both the newspaper comic strip and the radio show and could be accommodated easily, but the demand for spectacle would be less easily fulfilled. Serials were low-budget additions to the theatrical bill and were increasingly under the same pressure as animated cartoons following the Paramount decrees, described earlier in this chapter. While color filming was used for big-budget feature films, Columbia's *Superman* serial was filmed in black and white, stripping the character of one of his most distinctive identifying features. Furthermore, the live special effects available to the production were unable to offer a credible version of Superman's trademark flying: instead the producers turned to animation to depict Superman's flight.[46] While these limitations did not constrain the immediate commercial appeal of the serial, with Columbia producing and releasing a second story, *Atom Man vs Superman*, in 1950, the serial would seem to have been considered mainly suitable for a juvenile audience ("a natural to attract the kiddies")[47] and being consigned to Saturday morning matinees, foreshadowing the associations of this window in the schedule seen in later television.[48]

With so many earlier adaptations demonstrating the commercial viability of the Superman character, it is no surprise that a television adaptation would follow as the new medium rose to popularity in the early 1950s. In contrast to those earlier adaptations, *The Adventures of Superman* (first transmitted in 1953) would take a more narrow approach to the adaptation process. While the show retained the same name as the radio show, it otherwise only selected elements of the franchise that fit the emerging understanding of television's properties. Production of *The Adventures of Superman*, initially by Flamingo Films, was located on the West Coast of America, away from the center of production of the comic books and radio show (as well as the live television industry), moving to Hollywood, the site of rapid growth in filmed programming for television. This allowed the show to draw on the pool of talent newly available as the studios divested their contract players, just as they had released the animators discussed in the first part of this chapter in the same period. These included the star of the show, George Reeves, who had previously been under contract at Warner Bros. and 20th Century Fox, as well as many of the other cast and crew.[49]

The result of this separation from the earlier adaptations was a show that played to the expectations of television as an intimate medium, with an emphasis on closeness to the actors and dialogue rather than action and spectacle. As director Tommy Carr suggested "the writing was better [than the film serial] on television, the acting was better, too; it was a more intimate medium."[50] This was particularly evident in a shift toward Clark Kent as lead protagonist, with Reeves allowing little distinction between the reporter and Superman beyond the costume. Reviews of the show highlighted this, *Variety* writing of the theatrically released first episodes that Superman was "limited . . . to some routine superhuman stunts" while praising Reeves as "effective in both roles" and discussing the plot in terms of Clark Kent as protagonist.[51] The lack of spectacle did nothing to harm the show's popularity and it was an immediate and lasting success, with six seasons produced and a seventh being planned at the time of George Reeves's untimely death in 1959. The show continued to be syndicated long after production had halted and its continued success was credited as influencing the decision to start producing *The New Adventures of Superman* cartoon series in 1965, with *Variety* calling it a "numbers powerhouse" able to rival *I Love Lucy* in its ability to attract viewers on the thirteenth or fourteenth airing.[52]

The animated cartoon series *The New Adventures of Superman* was a product and logical extension of the two developments discussed in this chapter, and in examining it more closely, both in production terms and the content of the show itself, we can further understand how science fiction offered a space for the negotiation of the medium specificity of both television and animation. At a production level it marked the culmination of the move away from studio-era theatrical animation to the Saturday morning slot aimed at children. As a further adaptation of the Superman mythology it had to contend with the character's duality and find a way to balance or negate the conflict between human domesticity and superhuman spectacle.

Filmation, the animation studio responsible for *The New Adventures of Superman*, was formed in 1962 by Lou Scheimer and Hal Sutherland. Both Scheimer and Sutherland had started working in animation in the 1950s, gaining experience at established studios: Sutherland at Disney, Scheimer at Warner Bros. Yet, as shown earlier in this chapter, the animation industry and Hollywood more generally was undergoing turbulent change and financial difficulty, and Scheimer and Sutherland were left moving between short-term positions on a variety of productions. They met while working at Larry Harmon's studio on *Bozo the Clown* cartoons, one of the earliest made-for-television animated cartoons, and Scheimer also gained experience at Hanna-Barbera working on *Ruff and Reddy*, giving the pair experience in the very different production process required for low-budget television animation.[53]

They formed Filmation to produce *Rod Rocket* (1962) a series of five-minute animated shorts that was widely syndicated. This show demonstrated the appeal of science fiction for the new market for animated cartoons on television, with each show including a "space fact" tied in with a fact sheet that could be used by sponsors for merchandising opportunities.[54] They were joined by Norm Prescott when they started producing an animated sequel to *The Wizard of Oz*, but fell into financial difficulties given the expense and risk of producing an animated feature.[55] It was at this point, in 1965, that Prescott was able to secure the contract to produce *The New Adventures of Superman* for National Periodical Publications, saving Filmation from closure.

The New Adventures of Superman was a cornerstone of the plan by the American television network CBS to dominate the Saturday morning window by programming an all-cartoon schedule, with science fiction animation pivotal to it, extending the success seen by that channel with their *Mighty Mouse Playhouse* since 1955. CBS had been relegated to third place in the Saturday morning ratings, which were led by ABC's *The Beatles* cartoon.[56] Fred Silverman, head of daytime programming at CBS, is credited with adopting an "all-animated" schedule that included not only *The New Adventures of Superman* but also *Space Ghost*, *Frankenstein Jr. & the Impossibles*, and *Underdog*, all of which had significant science fiction elements; the schedule also included adapted theatrical cartoons in both *Road Runner* and *Tom and Jerry*.[57]

The patterns described in the earlier discussion of Saturday morning cartoons are clearly evident here. Advertising costs for this time period were far cheaper than prime time: per-minute rates for *The New Adventures of Superman* were $8,100 in 1967, the highest rate seen for Saturday morning shows, which more commonly attracted $4,000–$6,000. This was still far lower than ABC's prime-time *Batman*, which could charge $40,000 a minute on Thursday evening, a typical advertising price for prime-time shows.[58] By targeting a purely juvenile audience the Saturday morning period could offer advertisers attractive value for their money.

The Filmation Superman show was very clearly aimed at this audience. The trade press repeatedly described the shows as "kidvid," "kiddie enticers," "Saturday kiddie fare," and "all for the tots."[59] This was an intentional strategy to attract advertisers looking to exclusively reach children, as one article, titled "Kids? Let Them Eat Candy" wrote, "The toy companies, the cereals, the bubble gum firms want big Nielsen numbers, and by trial and error, the big networks have learned that the big Nielsen numbers can be delivered by cartoons."[60] This 1966 article estimated the advertising income of Saturday morning at $25–$30 million, by 1969 this had risen to $75–$90 million thanks to the slot providing a "demographically pure audience for advertisers with products to sell to the kiddies."[61]

The New Adventures of Superman was both designed to fit this definition of what television animation was and also one of the leading programs that reinforced it. While earlier adaptations of theatrical cartoons for television had little impact on the cartoons themselves, despite the redefinition of animation their scheduling practices brought about, the animated content created for television animation was greatly shaped by its intended exhibition medium. *The New Adventures of Superman* was initially designed to fit into a half-hour slot, with two seven-minute self-contained Superman episodes bookending a Superboy cartoon also produced by Filmation. The seven-minute length was considered well suited to television, as it did not require the episodes to be interrupted for advertising breaks, indicating the influence the structural flow of television could have on the design of programs; this format was also better suited to adapting comic book–length stories.[62] Competition increased sharply between 1966 and 1969, so in 1967 CBS extended *The New Adventures of Superman* to become *The Superman-Aquaman Hour of Adventure*, using the popularity of Superman to sustain a whole hour of programming.[63] As a result the Superman cartoon's third season introduced double-length episodes with a cliff-hanger break, a further attempt to maintain viewers' loyalty and keep them from switching to other networks with their own cartoon-heavy Saturday morning schedules.

A second fundamental change necessitated by the television medium was the budget available for production of cartoons. As part of its move to dominate the Saturday morning schedule, CBS was reported to have spent $8 million, with the Superman cartoons being a central part of this expenditure.[64] While a significant investment (Disney's seventy-five-minute feature *Sleeping Beauty* [1959] was reported to have cost $6 million),[65] this budget was intended to provide two years of programming for a weekly five-hour window between 8 AM and 1 PM, extended to 2 PM in later schedules.[66] Even with the intention that shows would be repeated four times within that two-year window, the result was that Filmation was given a budget of $36,000 per half-hour Superman show.[67] In comparison, Paul Terry suggested that his seven-minute theatrical shorts in the 1950s might have a budget of $45,000 to $50,000, and MGM *Tom and Jerry* cartoons could reach costs of $60,000 per short.[68] Even Hanna-Barbera's made-for-television animation was produced at a cost of $45,000 per half-hour show. The result was the need for Filmation to embrace every cost-cutting measure available, especially the reuse of animated sequences, heavy reliance on all elements of the soundtrack to fill in events and selective animation of movement. These characteristics led some commentators, still wedded to older definitions of animation, to dismiss these shows as "illustrated radio," as celebrated Warner Bros. director Chuck Jones did.[69] Yet the repeated animation served as a visual equivalent to Superman's catchphrases in the radio show, which also appeared in the television cartoons, a familiar presence that anchored the viewers' attachment to the show.

The intention here is not to reverse the valorization of "golden age" character animation or the denigration of later limited animation, more simply to note that *The New Adventures of Superman* and the other made-for-television shows offered a moment at which definitions of television and animation were undergoing considerable change. That evaluations of art forms often rest on their perceived adherence and exploitation of medium-specific qualities, as discussed in the introduction of this book, meant that judgments of these new shows depended upon the acceptance of the reviewer of those new definitions of what animation was on television. Put more simply, reviewers who embraced the new paradigm of animation on television as exclusively for a juvenile audience and embedded in the adaptation of a diverse range of sources praised *The New Adventures of Superman*; those committed to animation's expression of character through movement, and who saw in such expression the "essence" of animation, found the new programs to be a corruption of the form's ideal.

Stripped of its evaluative element, Jones's assessment of limited animation is perceptive: in present-day terms we might label it augmented radio. Thanks to their shared heritage and commonality as transmission media, television and radio were seen as having shared capacities and limitations that are evident in the choices made in *The New Adventures of Superman*. Again crediting Fred Silverman of CBS with the redefinition of animation on television, a 1967 profile of Filmation suggests Superman arose when Silverman "remembering the success of the old-time afternoon radio adventures . . . devised a similar formula of adventure block programming for his network on Saturday mornings." Like the Fleischer adaptation before it, Filmation hired the cast of the 1940s radio broadcasts to provide voices for its cartoon show, including the return of Bud Collyer as Superman, replicating his trademark shift in tone during the catchphrase "this looks like a job *for Superman*."[70] The show was also executive-produced by Allen Ducovny, who had produced the original radio show.[71]

Despite these close links, an animated television adaptation of Superman carried different expectations from the radio show or live-action adaptations. On the one hand animation, with its laborious production process, was not associated with the immediacy of a live radio broadcast, even with the aid of the radio voice actors. Equally, the sense of intimacy achieved in the George Reeves *Adventures of Superman* television show, or even the character animation in the Fleischer's Superman cartoons, was challenging to achieve with the limited animation required by the economic constraints of the television show. Rotoscoping was again used, as it had been for the Fleischer's adaptation, giving the show a more photorealistic look to provide some human appeal to the characters.[72] In addition to offering a greater connection with the otherwise limited character models, however, the use of rotoscoping also pushed *The New Adventures of Superman* toward the adaptation of qualities normally associated with another medium, live-action film.

The trade paper *Broadcasting* suggested as much in a profile in 1967, stating that Filmation's "rise to prominence is pegged to the new animation, the rendering of live-action adventure characters in animated form."[73] The move toward a live-action model, and particularly the quality of spectacle associated with that medium, was further in evidence by the writing process. The show was initially created by writers provided by National Periodicals who had worked on the comic book versions of Superman, and later Filmation started hiring live-action scriptwriters to create their shows.[74] The Disney model, which had become dominant in the animation industry, had seen animators skilled in developing a story, called "storymen," devise the narrative of an animation film or short by creating a storyboard as the primary planning tool.[75] In contrast, Filmation started to use a live-action model, creating a written script with dialogue and scene breakdown before the storyboard process began.[76] Here is further evidence of the hybridization of film and animation to produce a new form of television show: Filmation's "storyboard men function as motion picture illustrators. They bridge the gap between live action and animation."[77]

One result of this shift was the introduction of spectacle into the show, at variance from radio and television models but aligned with cinema. A 1967 account of the Saturday morning programs, including *The New Adventures of Superman*, describes them as "a visual and audio experience of shattering proportions" indicating the spectacle and scale offered, rivaling the big screen experience.[78] Animation even offered to go beyond what was possible in live-action cinema; as Lou Scheimer argued, "we could do things that they couldn't do in live action, which is why we had so many more super-villains and aliens and super-powers, but there were still limitations because of the time we had and the budget."[79]

Spectacle was also provided by the opportunities of color, growing in importance to television in this period. Just as Disney had been among the first companies to demonstrate the potential of color for cinema in the early 1930s, having signed an exclusive contract with Technicolor,[80] animation was at the forefront of the expansion of color television in the mid-1960s. Trade press reports consistently emphasized the role of color, with one account detailing CBS scheduling "four and a half hours straight of tint cartoons"[81] and another describing "the color animation mill"[82] and a third the "livid tint" of Saturday morning cartoons.[83] In these accounts color was an essential component of the programs' economic success, providing a spectacular eye-catching aesthetic, but they also express a degree of what David Batchelor calls "chromophobia," a sense that color is in some way excessive, particularly in destabilizing medium specificities of television such as realism and intimacy.[84] Color of course also linked the cartoon television show with the comic book version of Superman, reinstating his iconic red, blue, and yellow costume. As described earlier, the Superman comic books had been heavily influenced by cinema, so it is highly fitting that the animated television show should return to its comic book heritage through its adaptation of cinematic spectacle. This should not be considered a return to "origins" or to misconstrue the series as "true to its source material" as other writers have suggested.[85] Rather the animated television Superman is a negotiation of all the adaptations that preceded it, emphasizing some characteristics and diminishing others in an attempt to construct a new model for television animation.

An analysis of the Superman episodes themselves also reveals the constant negotiation of the competing medium definitions that the show navigates to accommodate the history of adaptations addressed above. Several of the overall choices the show makes are indicative of its approach. In contrast to the live-action television series starring George Reeves, *The New Adventures of Superman* is heavily dominated by Superman, with Clark Kent serving primarily as an entry point to an emergency requiring Superman and a point of closure for that narrative. The show frequently ends with Kent making a veiled reference to his being Superman and winking to the audience, who share the knowledge of his true identity. In this adaptation Clark Kent is thus merely a disguise rather than an integral part of the Superman persona or a fully fledged alter-ego. Where the live-action television show had emphasized Clark Kent and therefore the intimate nature of the television drama driven by dialogue, the Filmation series' focus on Superman leads to spectacular narratives.

The spectacular nature of the narrative structure of *The New Adventures of Superman* points toward the influence of both cinematic and comic book models—that is, models of transcription and reproduction—yet the plots themselves and their method of presentation remain committed to transmission and immediacy, qualities more commonly associated with radio and television. A recurring example of this is Jimmy Olsen's "Superman Signal Watch." Olsen is the young reporter for the *Daily Planet* who works alongside Clark Kent and Lois Lane and frequently finds himself in peril when seeking out news stories. Olsen has a wrist watch that, when activated, broadcasts an inaudible signal that Superman can detect and use to locate Olsen to save him. This device is repeatedly used in episodes, including "The Robot of Riga," "Invisible Raiders," and "The Return of Brainiac." Equally, in episodes such as "The Fire Phantom," "Luthor Strikes Again," "Luthor's Loco Looking Glass" and "Luthor's Lethal Laser" Jimmy is unable to use the watch and because of that is pushed into further trouble. The watch allows events occurring simultaneously in another location to be brought to Superman's attention, creating a connection between the two locations, just as transmission media, such as television and radio, did. This serves to create a sense of immediacy about the stories, a sense that they are occurring in real time. The restrictions of the show, both budgetary and in terms of screen time, did not allow for the methods live-action films might use to achieve a similar effect, such as lengthy establishing shots and expository scenes. Rather *The New Adventures of Superman* advances to the spectacular action sequences rapidly, and Jimmy's watch serves to rapidly establish the connection between cross-cut sequences.

A similar effect is seen in the repeated use of television and radio transmission technology to establish the peril Superman must resolve in order to bring each episode to a close. In "The Force Phantom" a giant monster made of energy attacks a "space base on the Asian mainland." *Daily Planet* editor Perry White and reporter Clark Kent are immediately informed in the next scene thanks to a teleprinter and are able to track the movement of the monster across the Middle East and Africa as it heads toward America. Transmission is also central to the aliens who unleashed the monster on Earth in this episode, as they monitor its progress via television screens in their spaceship orbiting the planet, and then watch on their screens as Superman battles the creature in space. Similar use of television screens to monitor activity in another location without complex crosscutting is seen in "The Threat of the Thrutans," "The Magnetic Monster," and "Luthor's Loco Looking Glass" among others.

An episode from the first season, titled "The Chimp Who Made It Big," best demonstrates in detail the way the constant interpolation of transmission media into the spectacular narratives of *The New Adventures of Superman* serves to give an additional sense of immediacy. In doing so it thereby negotiates an adaptation of both live-action cinematic qualities and those associated with television. It also serves to demonstrate the role science fiction plays in anchoring the fantastical elements of the show in the factual events of the 1960s, in this case the exploration of space. The episode begins with a sequence in space showing a craft that closely resembles the Mercury and Gemini capsules that would have been familiar to audiences from news coverage of American space missions. Inside the capsule is a chimpanzee named Toto, another detail that would recall the factual events of an early Mercury mission in 1961, in which a chimpanzee named Ham was used to test the process of putting a human astronaut into space. Accompanying these images is a voice-over identifying the "Venus space probe capsule" and describing the collision of a meteor and asteroid that releases "waves of radioactivity." The combination of the factual elements and the authoritative, documentary-style narrator lends the events an initial credibility. The episode then takes a turn toward the spectacular and fantastical, as Superman first fights his way through an asteroid field and then saves the capsule from burning up on reentry into the Earth's atmosphere.

As Superman returns the capsule to the NASA base, Lois Lane takes up the narration of the events on screen in the guise of a television reporter, reporting "live" from the top of an outside broadcast van with a television camera transmitting events. Lois addresses the audience directly and describes the crew opening the hatch of the space capsule, only to find the chimpanzee has been mutated by the radiation it was exposed to in space and transformed into a King Kong–like ape. Lois's narration serves not only to reinforce our understanding of the crudely animated images but also provides them with a sense of liveness, a simulation of real-life commentators that would be used to describe and interpret any live television broadcast. The episode then takes a further turn back toward cinematic spectacle as Toto abducts Lois just as Kong abducts Ann Darrow in the famous film. Superman, of course, is able to use his superhuman strength to save Lois and restore order. The conclusion of the narrative is told in a single shot of a newspaper with two panels showing the before and after effects of the radiation wearing off Toto. This shot evokes a further medium, newsprint, and especially the comic strips that served as a further source for this adaptation of the Superman myth, demonstrating the strength of this form to tell a story with just two frames. Thus in a seven-minute episode "The Chimp Who Made It Big" is able to utilize a science-factual documentary mode to introduce the story, evoke the live transmissions of television and radio to give the action on screen an immediacy associated with those media, depict spectacular and fantastical events that adapt one of cinema's most celebrated original works, *King Kong* (1933), and pay homage to the form of the newspaper comic strip that offered one of the earliest adaptations of Superman.

A further medium, radio transmission, is also central to many episodes of *The New Adventures of Superman.* "The Deadly Icebergs" features a mysterious group of hooded men who first use a radio transmitter to misdirect a boat toward some icebergs and then monitor its progress in real time via radar. In "The Ape Army of the Amazon" the titular army is controlled by a radio transmitter on the back of an evil scientist, while in "The Deadly Dish" Lex Luthor builds a "kryptonic transmittor" that is able to weaken Superman by broadcasting the effects of kryptonite from afar.

As these examples suggest, in addition to regularly depicting the use of radio and television technologies as they were defined in the 1960s when the show was produced, the episodes also use the capabilities of the science fiction genre to envisage new functions that suggest these technological media can change over time, with a consequent change to their medium-specific qualities. In "Merlin's Magic Marbles" Lex Luthor is able to communicate across both time and space with Merlin, King Arthur's wizard, via a television device that also allows two-way visual and aural communication. In "A.P.E. Strikes Again" Professor Noble has invented a "trouble televisor" that can look into the future, allowing him to watch for trouble and crime before it happens and alert Superman via a radio alarm signal to prevent the crime.

The most detailed exploration of these themes is in the episode "The Image Maker," in which another mad professor, here named Leo Nula, is seeking revenge against Lois Lane, whose reporting sent him to prison. The episode begins with Nula looking at a newspaper featuring a photograph of Lois Lane, to which Nula states "very pretty photograph Miss Lane. Soon that's all there'll be to remember you by," not only offering a threat that drives the narrative of the episode, but also alluding to the death-mask qualities that have often been considered an essential quality of photographic reproduction.[86] Nula has developed a form of 4D cinema that he promises "will revolutionize movies." The device he shows Lois and Clark Kent is an adaptation of aspects of both television and cinema, combined with science fiction embellishments that extend it beyond either medium. Nula emphasizes that the pictures are projected from behind the screen, an arrangement not impossible for theatrical presentations but atypical for cinema, whereas television had commonly placed the apparatus and light source in front of the viewer. As Nula begins the demonstration of his device the screen flickers

Figure 4.1. ***The New Adventures of Superman*** **episode "The Image Maker." Professor Nula demonstrates his new hybrid of television and cinema.**

and produces static noises, another quality associated with television broadcasts, and the professor, having just left the room, is seemingly being broadcast live onto the screen and directly addresses the audience. Yet the screen Lois and Clark watch him on is of a theatrical scale, enlarging Nula so that he becomes an intimidating presence looming over his viewers. Nula then demonstrates the futuristic innovation he has created, an extension into the fourth dimension by which he can throw an apple from the screen, an ability Lois describes as "fantastic" and one that would seem to combine the intimacy and immediacy of live television with the spectacle and scale of cinema. Clark, on the other hand, is concerned that "in the wrong hands this could be a dangerous device," which not only signals to the audience the next events in this cartoon's narrative, but also expresses unease at the breakdown of medium boundaries.

Clark is, of course, prescient in his unease, and as he leaves to change into his Superman guise a huge monster appears on screen, comes into the viewing space and abducts Lois. Superman attempts to approach the screen and save her but she is now, in the professor's words, "just an image on film," again implying an absence inherent in photographic media, and suggests "only I, Professor Nula, can restore her back to flesh and blood." Superman leaps at the screen like the "rube" in Edison's 1902 *Uncle Josh at the Moving Picture Show* seemingly expecting to enter the image, only to crash through and find only the mechanism creating the illusion. The story is resolved with a science fiction rescue, as Nula attempts to launch Lois into orbit in space and Superman rescues her and captures Nula. The typical epilogue again points to a tension between television and cinema, as Clark suggests that Lois is "pretty enough to be in the movies," to which Lois responds with an angry "movies?" and throws a book at Clark. It seems this Lois is more comfortable with television than cinema, or any hybrid of the two.

We have already seen that the industrial and institutional context of *The New Adventures of Superman* used this science fiction show as a basis for redefining the relationship between animation and television. Analysis of episodes of *The New Adventures of Superman* further demonstrates the way the show's stories were also actively negotiating different medium-specific qualities, and using science fiction to explore new, hybrid forms of media that suggest an awareness of the historically contingent nature of medium specificity and the possibility of redefining them further in the future.

The redefinition of animation for television by *The New Adventures of Superman* was embraced immediately by audiences and the show became a vanguard for this type of programming, initiating a host of science fiction imitators, especially comic book superhero adaptations including ABC's *Spider-Man* (1967) and NBC's *Birdman and the Galaxy Trio* (1967), and confirming the association of Saturday morning schedules with cartoons and a juvenile audience. CBS leapt from third place in the network ratings to first, displacing ABC, whose hugely popular *The Beatles* cartoon had been unassailable until this point. *The New Adventures of Superman* and another CBS show, *Frankenstein Jr. and the Impossibles* (a Hanna-Barbera production), were described as the "smash successes" of the "runaway" all-cartoon schedule.[87] Their success with this format led the other networks, ABC and NBC, to adopt the same programming strategy, NBC even going so far as to hire Superman creator Jerry Siegel to write for the show *Cool McCool* (1966), which was devised by Batman creator Bob Kane and scheduled in direct competition with *The New Adventures of Superman* in the same time slot.[88] *The New Adventures of Superman* was also exported internationally as far afield as Japan, Puerto Rico, and nine Latin American countries.[89]

The growth of these shows also brought increased scrutiny upon them, especially as the interest of advertisers, targeting a specifically juvenile audience through them, demonstrated the strong influence and effect these shows could have upon their impressionable audience. At first criticism was directed at the production values of the shows, with *Variety* describing the "endurance run [of] . . . cartoon grind" and predicting "possible blindness for any kid who watches it all."[90] In part this criticism reflected a valuation based on medium specificity. A 1967 *Variety* review describes the "striking contemporary color animation" of *The Lone Ranger* and *The New Adventures of Superman*, which it likens to "a closeup view of an H-bomb attack," contrasting them to recycled "golden era" theatrical cartoons, which are described more favorably as "Bunny classics . . . attractively presented."[91] Yet this review also reveals a more specific concern than simply the changing definitions of animation, condemning the Superman cartoon and its peers as "loaded with dark violence" and describing an episode of Superman "that would spook adult viewers let alone the scapes."[92]

The comic book and live-action television series adaptations of Superman had undergone similar scrutiny a decade earlier when Fredric Wertham's *Seduction of the Innocent*: *The Influence of Comic Books on Today's Youth* had been published, resulting in the Comics Code Authority being established and the television show toning down action sequences such as fight scenes.[93] Television animated cartoons would not allay fears so easily amid the complex cultural and political upheaval of 1968, including the country's involvement in Vietnam, and the assassinations of Martin Luther King and Robert F. Kennedy. On June 10th, four days after the Kennedy assassination, President Lyndon Johnson posed the following question: "Are the seeds of violence nurtured through the public's airwaves, the screens of neighborhood theaters, the news media, and other forms of communication from our leaders that reach the family and reach our young?" He called for radio and television heads as well as a new federal commission to scrutinize the media's role in cultivating a violent culture.[94] Johnson's accusation was taken up by pressure groups lobbying television stations, with Saturday morning cartoons a primary target. A 1968 issue of *Women's Wear Daily* directly criticized *The Superman-Aquaman Hour of Adventure*, among other shows, for its violence, while also reporting on the discontinuation of sales of toy guns at Sears, Roebuck, the type of toy that was heavily promoted during the advertisement breaks in the Saturday morning cartoons.[95]

One response was an increasing call for educational programming for children, with the launch of *Sesame Street* in 1969 being the most prominent result.[96] A further result was the television networks' decision to either soften the content of shows or withdraw them completely. *Variety* reported a "general cutback on violence" and this was attributed to NBC's success in the ratings with *Banana Splits Adventure Hour*, overtaking *Superman*, against which it was scheduled.[97] This scrutiny of children's television continued into the early 1970s, with pressure groups such as Action for Children's Television and the National Association for Better Broadcasting challenging station programming and succeeding in having shows broadcast with caution warnings or removed from station schedules completely. Superman cartoons, along with other science fiction comic book superhero shows such as *Spider-Man*, *Super Six*, and *Gigantor*, were among those targeted for removal from the airwaves.[98] Superman was too popular and successful a character to disappear completely, but in 1973 Hanna-Barbera took over the rights to the character and recast him in the show *Super Friends* alongside Batman, Wonder Woman, and Aquaman "in adventures that teach socially useful lessons."[99]

Filmation was equally able to survive this further change to the constraints on television animation: by 1968 they were considered, along with Hanna-Barbera, to "pretty well dominate kidvid cartoon production on Saturday and Sunday mornings" with six shows being aired and more in production.[100] They remained committed to science fiction as a subject, seen in both *Journey to the Center of the Earth* (1967) and *Fantastic Voyage* (1968). In both cases these shows indicated the further importance of the adaptation of live-action qualities into television animation, as both series were directly based on theatrical feature films from 20th Century Fox.[101] *Journey to the Center of the Earth* was also, of course, based on the work of Jules Verne, lending the show an association with classic literature that would appease critics of children's television. Such an approach would prove a growing trend in the 1970s, as the case study of adaptations of Verne's *Twenty Thousand Leagues Under the Sea* (1869–1870) in the final part of this chapter will demonstrate.

TWENTY THOUSAND LEAGUES UNDER THE SEA AND THE HETEROGENEITY OF ANIMATED TELEVISION ADAPTATIONS

There are a number of explanations why Jules Verne's *Twenty Thousand Leagues Under the Sea* (*Vingt mille lieues sous les mers*),[102] originally published in serialized form in 1869–1870, should have seen a flurry of animated adaptations one hundred years later. The adaptation of a "classic" work of literature clearly had considerable appeal for producers and broadcasters to counteract the attacks on Saturday morning cartoons for their violence and lack of educational value, discussed in the previous section, but this does not explain the appeal of this particular work as the range of texts available fitting this purpose remained huge. The most immediate stimulus must be considered the rerelease of Disney's theatrical film adaptation *20,000 Leagues Under the Sea*. Originally released in 1954, this film had been a huge success, and its rerelease in 1971 was equally successful and brought renewed attention to its literary source.[103] Yet beyond these contextual reasons, the story and its version of science fiction would themselves seem to be particularly well suited to the interrogation of medium-specific qualities and as a consequence had been chosen for adaptation frequently since the novel's first appearance.

Even more than Superman, *Twenty Thousand Leagues Under the Sea* might be considered a source with no master text and as a consequence all versions may be understood as a form of adaptation. William Butcher in an introduction to his recent translation describes how this work was published in the original French language as both a serial and a single volume and Verne's publisher, Hetzel, exercized considerable influence over the material, introducing variations between all editions.[104] Furthermore, in the English language the novel was not only subject to the inevitable vagaries of changing translation practices, but was also frequently abridged, with up to 22 percent of the French editions removed, and it has been presented in hundreds of editions with a number of translations.[105] Of particular significance for the current study, the role of illustrations in visualizing the work might also be considered to undermine the notion of an original source for adaptation. Illustrations by Riou and de Neuville were added to an 1871 French edition and were subsequently included in some English-language editions but not others. An 1875 edition published by George M. Smith & Co in Boston included some of the Riou and de Neuville illustrations, but it was adapted to remove French-language elements from the images and typeset differently.[106] An 1892 British edition published by Ward, Lock & Co does not feature the illustrations but does feature a color frontispiece by a different artist, while the most recent Oxford World's Classic edition has no illustrations.[107]

Beyond these matters of textual difference, Verne's story is itself concerned with the limitations and difficulties of adaptation, especially in relation to science fiction. The story concerns the abduction of Professor Aronnax, an expert in marine biology; his domestic servant, Conseil; and Ned Land, a harpooner. The trio are rescued from the sea by Captain Nemo after he has attacked their ship with his technologically advanced submarine *Nautilus*, however Nemo then keeps them captive while taking them on an episodic journey through the seas of the world. The novel is narrated by Aronnax in the first person, and as with any story in this form immediately foregrounds the observer's role in constructing the events presented. In narratological terms a first-person account dramatizes the difficulty, if not impossibility, of distinguishing story/fabula/discourse from plot/syuzhet/histoire. Such a position suggests that a story can only be experienced in its original expression, and thus is highly pertinent to the challenges of adaptation, of translating a story experienced in one medium into another. This is particularly the case in *Twenty Thousand Leagues Under the Sea* due to its science fiction elements, presenting as it does possible but improbable technologies and natural phenomena. Nemo's technological medium, the *Nautilus*, transports Aronnax and his colleagues into new worlds and experiences, offering a narrative they cannot imagine in any other form.

Aronnax is a man of science and committed to an attempt to present these experiences objectively.[108] The temporal flow of the novel is frequently interrupted by detailed taxonomic accounts of the marine life seen from the windows of the *Nautilus* as well as the futuristic technology of Nemo's inventions in the chapters "All by Electricity" and "A Few Figures."[109] Yet Aronnax's ability to fully describe the story is flawed. While some gaps in his account, such as his failure to determine and describe Nemo's motivation, may be seen as mere character flaws, these gaps can also be seen as a meditation on the capacities of different media and the ability to translate their contents into others. On a number of occasions Aronnax questions the ability of language to communicate his experience. In one such passage he writes that he is "incapable of depicting the unique effects of the liquid element, how could a pen begin to portray them?"[110]

Similarly, in attempting to describe a squid attack upon the *Nautilus* and its passengers in which one of the crew dies, Aronnax writes,

> None of us will ever forget that terrible scene of 20 April. I wrote about it under the imprint of powerful emotions. Since then, I have re-examined the narrative and read it to Conseil and Ned. They found it factually correct, but too pallid. To paint such a picture would take the pen of the most illustrious of our poets.[111]

This passage is particularly revealing, as it is the squid attack, as will be seen, that would prove to be a crucible for later adaptations and their use of the story to forge new definitions for its expression in their chosen medium.

Twenty Thousand Leagues Under the Sea would prove a popular source for adaptation following its publication. In each case the adaptation would seem to raise questions about the ability of the chosen medium to translate the novel or seem to have been embraced by artists looking to expand the expressive range of their chosen medium. In 1890 Professor Pepper, best remembered today for giving his name to the "Pepper's Ghost" stage effect, presented an entertainment based on the novel, in which he recounted the plot of the novel illustrated by magic lantern slides depicting key scenes.[112] Georges Méliès, famed for his innovations in the field of trick films and the creation of new techniques for them, produced a loose parody of the story in 1907 titled *Under the Seas* (*Deux cent mille lieues sous les mers*). In 1924–1925 Verne's novel was used as the basis for an interactive installation at the British Empire Exhibition held at Wembley near London, offering an early attempt to allow visitors to experience the undersea world rather than have it described, something repeated in a number of later Disney theme park experiences.[113] In 1916 Universal produced a feature film based on the novel, directed by Stuart Paton and starring Alan Holubar as Captain Nemo. This costly adaptation was heavily dependent upon the use of a "cinematographic submarine vessel" to film scenes underwater, which technological innovation was emphasized in advertising and publicity for the film.[114] In 1952 the story was adapted for television in two episodes of ABC's *Tales of Tomorrow* at a time when Disney was known to be preparing a feature film adaptation. A *Variety* story announcing this seemed to question the capacity of a live-action television show to capture the fundamentals of Verne's story: "Interestingly, while 'Leagues' will be done with a combination of animation and live action by Walt Disney in Hollywood, producers of 'Tomorrow' feel that they can achieve all the undersea effects live on TV."[115] Disney itself was under scrutiny for its adaptation, released in 1955, with the *Times* suggesting, "To transfer the book to the screen will tax even the ingenuity of Mr. Disney and his helpers" while the film was still in production.[116]

The Disney adaptation warrants further attention, not only because it is the most proximal cause of the choice of this story for the 1970s animated adaptations, but also because of Disney's dominant association with animation. As suggested previously, Disney was heavily associated with animation in the "golden age" with both its short films and features achieving huge financial and critical success. The challenges of filming Verne's novel might seem to point naturally to the use of animation as the most suitable form for adaptation, and while *20,000 Leagues Under the Sea* may have initially been considered as an animated feature, the finished film was almost exclusively made using live-action and physical effects.[117] Some animation was produced with a view to using this form to depict marine life but this material was excised from the finished film. This certainly was not due to the quality of the animation, which has since been released and exhibits the skill of the Disney animators at this point, evoking the beautifully intricate movement of their earlier work in *Fantasia* (1940) as well as the abstract animation of Oskar Fischinger.[118]

The exclusion of Disney's core activity from the film instead reflects changing definitions of theatrical film and animation, alongside the growing importance of television as both an influence and new medium for exploitation. The simplistic connection between widescreen processes and television has been overstated and is rooted in the more complex changes in Hollywood described earlier; nevertheless, differentiating theatrical film from television was a key factor in Disney's first adoption of the Cinemascope technology for *20,000 Leagues Under the Sea.*[119] Equally we have seen in an earlier chapter the growing importance of television to Disney as both a source of income and outlet for their productions, as part of Disney's business diversification, which also included the growth of live-action features and the building of their first theme park. In terms of the media texts themselves, J. P. Telotte sees this shift as one of "convergence," suggesting medium-specific concerns were marginalized by the creation of transmedial stories, an approach that has clearly become dominant in contemporary Hollywood production.[120] It is certainly the case that *20,000 Leagues Under the Sea* as a feature film was heavily supported by both the "Operation Undersea" episode of the *Disneyland* television series, which recounted the making of the film, and the theme park attractions based on the film, both the initial display of the sets in the California Disneyland park and the ride opened in 1971 at the new Walt Disney World resort in Florida, coinciding with the rerelease of the film.[121] Yet these different expressions of the story remained embedded in an understanding of their supporting media, definitions that were being reconfigured by the institutional changes they were undergoing in the 1950s: the feature film increasingly concerned with the spectacular, television as documenting and intimate, and theme parks as experiential.

There are many examples of the way the Disney feature adapts Verne's science fiction to fit the increasingly spectacular model of theatrical film, but the most notable is its handling of the squid attack. In Verne's novel the chapter simply titled "Squid" is very brief and is typical of Aronnax's first-person account, giving considerable space to a discussion between himself, Ned Land, and Conseil about the taxonomy—and their scientific knowledge of—these creatures. The attack upon the *Nautilus* by numerous giant squid is described in only the last four pages of this short ten-page chapter.[122] In contrast, the Disney adaptation makes this attack the climax of the film and considerable effort was made to create a spectacular and engaging battle. The squid attack was initially filmed in daylight conditions, but on viewing the footage director Richard Fleischer and Walt Disney felt it did not achieve the required effect and the mechanical nature of the creature and the harnesses lifting the actors were evident. As a consequence the sequence was reshot in simulated nighttime and storm conditions, serving not only to mask the means of production, but also to further add to the dramatic effect of the sequence.[123] Typical of its approach to the adaptation, Disney excludes the scientific detail present throughout Verne's novel in favor of the brief, fantastic action segments.

The rerelease of Disney's film in 1971 thus provided several complementary motivations for *Twenty Thousand Leagues Under the Sea* as a source for television animation adaptation in the early 1970s. Its repeated commercial success demonstrated a ready-made audience; likewise, Disney's emphasis of the spectacular elements of the story was in keeping with the new paradigm for science fiction animation seen in *The New Adventures of Superman* and the other shows of the late 1960s. Yet, just as Disney was using Verne's science fiction to explore a redefinition of cinema in the aftermath of the Paramount decision and the arrival of television, so animated versions of this story were not simply intent on slavishly copying the Disney model, but saw in this work a way to renegotiate the role of animation on television in the light of institutional and technological changes. In particular, adapting classic works of literature offered protection from the criticisms aimed at shows based on comic books, while science fiction offered a suitable genre for exploring new distribution approaches and technologies.

The first animated television version of Verne's novel to be considered here is the one produced by Rankin/Bass in 1972. Rankin/Bass had been formed in the 1950s as Videocraft International, dedicated to producing commercials. It was founded by former ABC art director Arthur Rankin and advertising agency employee Jules Bass, and expanded into the production of television series in 1961.[124] Like Hanna-Barbera and Filmation, Rankin/Bass was able to develop animation processes that were economically viable for television shows: a key part of their cost control was subcontracting the animation work to Japanese companies.[125] The company initially adopted a stop-motion puppet animation technique they called "Animagic," which was used for a sequence of seasonal television specials, including *Rudolph the Red-Nosed Reindeer* (1964) and *Santa Claus Is Coming to Town* (1970), which would establish the Rankin/Bass name and have considerable longevity, being rerun annually and becoming Christmas institutions in the United States.[126] The company also started using cel animation techniques, such as that used on their 1969 Christmas special *Frosty the Snowman* as well as their version of *Twenty Thousand Leagues Under the Sea*.[127]

Rankin/Bass had largely avoided the comic book science fiction trend of the 1960s, the exception being their series *The King Kong Show* (1966) which featured *Tom of T.H.U.M.B.* a US intelligence agent who had been shrunk by a laser beam.[128] Yet they were still subject to the same shift in critical opinion discussed in relation to Superman. For example their 1968 special *The Mouse on the Mayflower*, which told the story of the first Thanksgiving through the eyes of a mouse, was derided by *Variety* as both "tasteless" and "saccharine" and most importantly "performed no service toward the education of the young."[129]

In response Rankin/Bass emphasized even further the prestige and educational value of their productions. Their 1972 show for ABC titled *Kid Power* offered didactic storylines and a diverse set of characters "reflecting most of the better values critics of kidvid would like to see utilized more."[130] Part of this move to appease critics was the launch of the *Festival of Family Classics*, which included their version of *Twenty Thousand Leagues Under the Sea*. In addition to responding to the changing critical climate for animation, *Festival of Family Classics* was also devised to respond to changing distribution patterns within television. In 1970 the Federal Communications Commission (FCC) instituted the Prime Time Access Rule.[131] This restricted the number of hours during the evening prime time period that could be programmed by the television networks, ABC, NBC, and CBS, on affiliated stations, with the intention of increasing competition and providing a greater role for independent productions.[132] The result was a growth in demand for family-oriented shows that could be broadcast in the newly available prime-time slots, and companies such as Rankin/Bass speculatively produced shows that could be syndicated for this purpose. *Festival of Family Classics*, including the *Twenty Thousand Leagues Under the Sea* episode, was thus aimed at addressing a further institutional change to television and repositioning animation as a family entertainment, rather than purely juvenile, and using the literary credentials of Verne's work and the other classic works of literature to distance itself from the maligned comic book adaptations.

To some extent, this further repositioning of animation is evident in the Rankin/Bass *20,000 Leagues Under the Sea*. The show's title sequence, shown on every episode, begins with a book opening to reveal the series title, *Festival of Family Classics*, whose name clearly emphasizes that this show is not exclusively for a juvenile audience. It continues with a library of books flying toward the screen before the characters from those books are brought to life all marching to a tune with the lyrics, "All the tales you know so well . . . each classic creation in new animation," the rhyme linking animation closely with the "classic" nature of the tales. The title sequence ends with a shot of Verne's novel, again emphasizing both the literary source of the episode, as well as placing Verne's work among other long-established works of children's fiction. The post-title sequence does not enter full animation, but instead presents the early accounts of the *Nautilus*'s sightings through a series of line drawings that evoke (but do not replicate) the illustrations that accompanied some versions of Verne's novel.

These attempts to distance the show from the Saturday morning cartoon model are also evident in some aspects of the narrative and animation. Most notable is the introduction of a partially anthropomorphized porpoise called Fifi that helps Aronnax and his companions escape danger several times. The introduction of this new character constitutes the most significant departure from Verne's novel, in an adaptation that otherwise makes considerable effort to include the many episodes of the source, including pearl diving, sunken ships, the Arabian passage, the South Pole, Atlantis, and the Sargasso Sea. Incongruous as Fifi is, her presence is indicative that Rankin/Bass's source is, at least partially, the work of Disney. Not only did the Disney live-action film similarly include a sea lion called Esmerelda to provide comic relief, but more importantly Fifi restores the association of animation with animal characters that had been dominant in the theatrical period but had been sidelined by human characters in television animation, including *The New Adventures of Superman*.

Despite these attempts to reposition television animation by associating it with classic literature and the longer tradition of theatrical cartoons and features, the Rankin/Bass adaptation is conflicted as it remains embedded within the new paradigm for television animation. Most of the *Festival of Family Classics* series was animated by the Japanese Mushi Studio, whose animation resembles 1950s Hollywood productions, such as UPA's *Mr. Magoo* series with its caricatured and cartoonish style. In contrast, *20,000 Leagues Under the Sea*'s animation was produced by Toru Hara, who would later work on Studio Ghibli productions, including *My Neighbor Totoro* (1989). As a result the episode has a strong comic book aesthetic that resembles the anime style associated with Japanese animation, with human characters that have a greater photorealistic appearance. Furthermore, this episode heavily emphasizes action and spectacle, continuing the appropriation from live-action cinema seen in *The New Adventures of Superman*.

As with the Disney adaptation, the octopus attack provides the clearest indicator of this spectacular approach.[133] The Rankin/Bass version of this sequence follows Disney's lead, offering a tense buildup as we see the octopus through the *Nautilus*'s portholes and then a battle between the crew and the marine creature that continues for nearly two minutes. Elements of the Disney depiction are repeated unchanged: the creature's tentacles reaching into the *Nautilus*, Ned Land being held captive by the creature, the crew hacking at multiple of the octopus's limbs with axes. Most notable is the rapid montage, resulting in around thirty shots for the sequence, and the highly dramatic music that recalls the Disney cinematic score dominated by an intrusive brass section.

The Rankin/Bass adaptation of Verne's work again demonstrates the complex and shifting negotiation of television animation. Verne's novel had clearly been chosen for its literary associations, placing it alongside other "classic" tales that were more commonly associated with animation, such as *Sleeping Beauty*, *Robin Hood*, *Snow White*, *Cinderella*, and *Alice in Wonderland*. This not only served to redirect criticisms of children's television but also established the family audience credentials necessary to exploit the new Prime Time Access Rules the series was targeted at. While the episode includes elements that would seem to return to earlier theatrical models of animation, it is ultimately dominated by an example of the spectacular, action-oriented narratives that had become dominant from the mid-1960s in television animation, led by *The New Adventures of Superman* and other science fiction series.

In contrast to the Rankin/Bass adaptation that attempted to find a balance between competing demands, Hanna-Barbera's 1973 adaptation moved more definitively to satisfying new expectations of animation and embracing new technological developments that were starting to reshape television. As discussed earlier, Hanna-Barbera had not only been innovators in the field of television animation but had remained one of the dominant studios for that form into the 1970s. The decision by them to adapt *Twenty Thousand Leagues Under the Sea* was motivated not only by the Prime Time Access Rule, but also the desire to exploit the new home format of "video home cartridges."[134] Hanna-Barbera understood this market to be aligned with educational needs and saw an adaptation of Verne's novel, alongside other familiar literary sources such as *Robin Hood* and *Oliver Twist*, as well suited to this aim.[135]

The show itself, produced in conjunction with Spanish animators at Estudios Filman, shows clear signs of this new direction.[136] The literary credentials of the show are again clearly highlighted by placing Verne's name on the title screen, as both Disney and Rankin/Bass had done, but the Hanna-Barbera adaptation also retains the wording of Verne's title, rather than spelling the *20,000* in numerals as did the preceding adaptations. While retaining a number of the spectacular events of the novel, this version is far more dialogue driven and treats these events very briefly in comparison to the earlier moving-image adaptations. The dialogue reintroduces elements of Aronnax's scientific account of the trip and Nemo's lectures.

The squid attack again proves the best gauge of the objectives of this adaptation. The Hanna-Barbera version uses the octopus attack not to produce an exciting, spectacular finale, but rather to induce a creeping dread upon the viewer that reinforces the educational message delivered by Captain Nemo. Nemo has taken Aronnax and Ned Land to the Sargasso Sea, which is littered with hundreds of wrecked ships, gathered here from all the oceans by the winds. Nemo proceeds to deliver a speech tinged with pacifist and environmentalist sentiments:

> Can't you understand what this desolation means to me? It represents all that I despise: the destroyers, the polluters. Look around Professor Aronnax. It will not happen in your day or mine, but the day will come when all the oceans of the world will be like this, unless we stop all those who knowingly seek to obliterate all that is beauty, all that is purity, all that is life.

During this speech shots of Nemo, Aronnax, and Land are intercut with shots of an octopus slowly creeping up the *Nautilus* behind them, accompanied by a slow unsettling woodwind score. The anticipation of the octopus attack underpins the troubling predictions Nemo is making for the future. Of course Nemo's warning is not to Ned or Aronnax, rather the viewer, for whom Nemo's future is their present day, given the historical setting of the show. What is a science fiction prediction from Verne and Nemo becomes a disconcerting impending reality for a late twentieth-century audience. When the octopus attack does finally occur it is completed perfunctorily in just over thirty seconds and only ten separate shots and a relatively subdued score. This approach is typical of the show as a whole, downplaying spectacular incidents and foregrounding an environmental message.

The Rankin/Bass adaptation of Verne's novel had allied science fiction with the spectacular, in keeping with both the Disney feature film and the most recent associations of television animation seen in *The New Adventures of Superman*, while the novel's literary credentials served to balance these with the implication of respectability. In contrast, for Hanna-Barbera the historic science fiction of Verne's novel offered a way to foreshadow present-day environmental concerns that upheld the increasing demand for children's television to have a didactic element that was reinforced by a new distribution technology. The appeal of Verne's work for this purpose was equally adopted in the Canadian series *The Undersea Adventures of Nemo* produced by Rainbow Animation.[137] Created in 1975 by a Toronto-based animation studio, the series only loosely adapted the novel, primarily using the Nemo name, its underwater connections, and a science fiction license for new technology. Episodes featured Captain Nemo, far more genial than seen in other versions, taking two young children undersea to explore the wonders of nature under the aegis of the "United World Organization" and carry out environmental missions while delivering educational sound bites on the marine life seen.

CONCLUSION

The evidence of the case studies presented here suggests that television as a medium, animation as a form, and science fiction as a genre, rather than having inherent definitions, are subject to historically and culturally specific constitution. This is not to suggest that these terms have no value. On the contrary, investigating the shifting definitions according to which these subjects were discussed and implemented as media texts at particular moments allows greater insight into the way they reflected and influenced the wider social context from which they derived.

NOTES

1. Barrier, *Hollywood Cartoons*.
2. Mittel, "The Great Saturday Morning Exile," 35–36.
3. *Broadcasting*, May 22, 1967, 66.
4. Maltin, *Of Mice and Magic*, 137.
5. Scivally, *Superman on Film, Television, Radio, and Broadway*, 9, 16.
6. *Boxoffice*, November 14, 1953, 8.
7. *Broadcasting*, April 11, 1955, 68.
8. *Variety*, December 28, 1955, 1.
9. *Billboard*, January 21, 1956, 17.
10. *Variety*, June 20, 1956, 23.
11. Mittel, "The Great Saturday Morning Exile."
12. *Billboard*, January 21, 1956, 17.
13. *Billboard*, May 12, 1956, 4.
14. *Billboard*, March 10, 1956, 14.
15. Mittel, "Great Saturday Morning Exile," 48–49.
16. *Variety*, November 21, 1956, 27.
17. *Broadcasting*, June 26, 1961, 98.
18. *Broadcasting*, April 3, 1967, 137; *Broadcasting*, May 29, 1967, 55. Mighty Mouse did not disappear, however, and he returned in the late 1970s in a newly animated series produced by Filmation, the studio responsible for *The New Adventures of Superman*. Scheimer, *Lou Scheimer*, 163–66.
19. *Variety*, June 20, 1956, 23.
20. *Broadcasting*, July 2, 1956, 14.
21. *Variety*, September 5, 1956, 21.
22. *Variety*, October 17, 1956, 32; *Billboard*, February 3, 1958, 11.
23. *Variety*, October 7, 1959, 42–45; *Broadcasting*, January 9, 1961, 76.
24. *Variety*, June 7, 1959, 38.
25. *Variety*, October 12, 1955, 4; *Billboard*, November 21, 1970, 16.
26. Mittel, " Great Saturday Morning Exile," 41.
27. *Variety*, February 9, 1949, 30.
28. Maltin, *Of Mice and Magic*, 300.
29. *Variety*, June 17, 1959, 27; *Variety*, May 24, 1961.
30. Daniels, *Superman*.
31. Scivally, *Superman on Film, Television, Radio, and Broadway*, 9.
32. Daniels, *Superman*, 11; Scivally, *Superman on Film, Television, Radio, and Broadway*, 8–9.
33. Scivally, *Superman on Film, Television, Radio, and Broadway*, 9.
34. "Superman Radio Episode List." *Superman Homepage*. http://www.supermanhomepage.com/radio/radio.php?topic=radio-episode-list.
35. Scivally, *Superman on Film, Television, Radio, and Broadway*, 15.
36. Ibid., 19.
37. *Boxoffice*, September 7, 1940, 30; *Variety*, September 4, 1940, 7; *Boxoffice*, October 5, 1940, 28.
38. Crafton, *Shadow of a Mouse*, 199–201.
39. J. P. Telotte discusses in more detail the competing demands the Fleischers' faced in adapting Superman and draws a similar parallel between these and the duality of Superman's persona: Telotte, "Man and Superman: The Fleischer Studio Negotiates the Real."

40. Fleischer, *Out of the Inkwell*, 105.
41. *Boxoffice*, October 5, 1940, 28; *Boxoffice*, August 23, 1941, 50.
42. *Variety*, January 7, 1942, 45.
43. Scivally, *Superman on Film, Television, Radio, and Broadway*, 26.
44. *Boxoffice*, July 17, 1948, B14; *Variety*, May 3, 1950, 7.
45. *Boxoffice*, January 17, 1948, 47.
46. Scivally, *Superman on Film, Television, Radio, and Broadway*, 36.
47. *Boxoffice*, July 17, 1948, B14.
48. *Boxoffice*, February 17, 1951, 42.
49. Scivally, *Superman on Film, Television, Radio, and Broadway*, 46–48.
50. Tommy Carr quoted in ibid., 49.
51. *Variety*, December 12, 1951, 6.
52. *Variety*, December 22, 1965, 27.
53. Scheimer, *Lou Scheimer*, 34–36.
54. *Broadcasting*, May 7, 1962, 83.
55. Scheimer, *Lou Scheimer*, 38–41.
56. *Variety*, December 29, 1965, 22; *Variety*, February 2, 1966, 34.
57. *Variety*, October 5, 1966, 25.
58. *Broadcasting*, February 6, 1967, 24.
59. *Variety*, December 22, 1965, 27; *Variety*, December 29, 1965, 22; *Variety*, October 5, 1966, 25; *Variety*, September 14, 1966, 31.
60. *Variety*, May 18, 1966, 25.
61. *Variety*, February 5, 1969, 39.
62. Scheimer, *Lou Scheimer*, 47.
63. *Back Stage*, June 2, 1967, 2.
64. *Variety*, October 5, 1966, 25.
65. Maltin, *The Disney Films*, 154.
66. *Broadcasting*, May 29, 1967, 55.
67. *Variety*, October 5, 1966, 25; Scheimer, *Lou Scheimer*, 45.
68. *Variety*, October 12, 1955, 4; Mittel, "The Great Saturday Morning Exile," 38.
69. Barrier and Spicer, "An Interview with Chuck Jones." Republished at http://www.michaelbarrier.com/Funnyworld/Jones/interview_chuck_jones.htm.
70. *Variety*, March 9, 1966, 26.
71. *Broadcasting*, January 31, 1966, 60.
72. Scivally, *Superman on Film, Television, Radio, and Broadway*, 72.
73. *Broadcasting*, May 22, 1967, 66.
74. *Variety*, May 24, 1967, 43.
75. Thomas and Johnston, *The Illusion of Life*, 376–83.
76. *Variety*, May 24, 1967, 43.
77. Ibid.
78. *Variety*, January 4, 1967, 82.
79. Scheimer, *Lou Scheimer*, 46.
80. Maltin, *Of Mice and Magic*, 39.
81. *Variety*, March 16, 1966, 34.
82. *Variety*, March 9, 1966, 35.
83. *Variety*, January 4, 1967, 82.
84. Batchelor, *Chromophobia*.
85. Perlmutter, *America Toons In*.
86. See, for example, the discussion in Mulvey, *Death 24x a Second*, 54–66.
87. *Variety*, September 14, 1966, 31.
88. *Variety*, January 4, 1967, 82; *Variety*, April 27, 1966, 35.
89. *Variety*, September 28, 1966, 45.
90. *Variety*, March 16, 1966, 34.
91. *Variety*, January 4, 1967, 82.
92. Ibid.
93. Scivally, *Superman on Film, Television, Radio, and Broadway*, 53, 56.

94. Johnson, "Remarks and Statement upon Signing Order Establishing the National Commission on the Causes and Prevention of Violence."
95. *Women's Wear Daily*, June 24, 1968, 18.
96. *Back Stage*, June 20, 1969, 7, August 8, 1969, 11.
97. *Variety*, February 5, 1969, 39, December 25, 1968, 26.
98. *Variety*, October 3, 1973, 20.
99. *Broadcasting*, April 16, 1973, 39.
100. *Variety*, June 19, 1968, 44.
101. *Broadcasting*, April 1, 1968, 114, May 6, 1968, 71.
102. I have retained the traditional, singular translation of Verne's title due to its familiarity and consistency with adaptations; however, as William Butcher's recent translation points out, a more accurate translation uses the plural "seas," which serves to emphasize that it is the distance circumnavigated that is referred to in the title, not the depth Captain Nemo's submarine dives to. Verne, *Twenty Thousand Leagues Under the Sea* [1998].
103. *Independent Film Journal*, May 13, 1971, 28; *Variety*, October 6, 1971, 3.
104. Butcher, "Introduction," xvi–xviii.
105. Ibid., x–xi; xxxii–xxxv.
106. Verne, *Twenty Thousand Leagues Under the Seas* [1875].
107. Verne, *Twenty Thousand Leagues Under the Seas* [1892]; Verne, *Twenty Thousand Leagues Under the Sea* [1998].
108. In this respect Verne's narrator shows striking parallels with the detached and anti-sensational writing of John Wyndham and his narrator discussed in chapter 5.
109. Verne, *Twenty Thousand Leagues Under the Sea* [1998], 76–88.
110. Ibid., 108.
111. Ibid., 348.
112. *Times*, October 3, 1890, 10.
113. *Times*, May 23, 1925, viii.
114. *Billboard*, July 22, 1916, 53.
115. *Variety*, January 9, 1952, 40.
116. *Times*, January 28, 1954, 11.
117. *Treasures of the Walt Disney Archives* exhibition, Museum of Science And Industry, Chicago; Leonard Maltin suggests some animation remains in the film to depict fishes in the *Nautilus*'s windows; if this is the case, it is very discreet. Maltin, *The Disney Films*, 122.
118. *20,000 Leagues Under the Sea*, Special Edition DVD, 2003.
119. Telotte, *The Mouse Machine*, 84–85.
120. Telotte, "Science Fiction as 'True-Life Adventure.'"
121. Taves, "The Making of *20,000 Leagues under the Sea*."
122. Verne, *Twenty Thousand Leagues under the Sea* [1998], 338–47.
123. Maltin, *The Disney Films*, 121–22.
124. *Broadcasting*, February 13, 1961, 72; *Variety*, March 15, 1961, 26; *Variety*, February 24, 1964, 85; Goldschmidt, *The Enchanted World of Rankin/Bass*.
125. *Broadcasting*, May 15, 1961, 105.
126. *Variety*, February 2, 1972, 51.
127. *Broadcasting*, February 13, 1961, 72.
128. Goldschmidt, *The Enchanted World of Rankin/Bass*, 97.
129. *Variety*, November 27, 1968, 42.
130. *Variety*, August 30, 1972, 40.
131. *Variety*, February 16, 1972, 35–38, March 15, 1972, 55.
132. *Broadcasting*, June 15, 1970, 40.
133. Verne's original French uses the term *poulpe*, which does not distinguish between squid and octopus. William Butcher suggests Verne intended *squid* and translates this archaic term as such, and in the Disney version the creature in the film both is called a squid and also resembles a squid rather than an octopus, yet Rankin/Bass use *octopus* in both dialogue and visual resemblance. Given the fantastical nature of the creature's attack its exact taxonomy is immaterial. Butcher, "Introduction," xxxvii.
134. *Billboard*, November 21, 1970, 16.

135. Ibid.
136. *Variety*, January 9, 1974, 88.
137. *Variety*, July 13, 1977, 51.

Chapter Five

Future Catastrophes

The Day of the Triffids *and* The Tripods

ESCHATOLOGICAL FANTASIES

John Wyndham was a British science fiction writer who has been described as the "master of the cozy catastrophe."[1] He had already been publishing in the American pulps when *The Day of the Triffids* (1951) was serialized in *Collier's*, a general-interest magazine that had built a reputation for investigatory journalism. *Collier's* had included some science fiction, though this tended to be more speculative than science, and had encouraged interest in space exploration in the early 1950s with illustrated articles by science writer Willy Ley and the artist Chesley Bonestell. Wyndham wrote about space flight, but he explained that for some reason the British public had not taken to it.[2] Brian Aldiss criticizes Wyndham's apparent lack of enthusiasm for space exploration by commenting that "he wrote about space flight, but, one feels, without any burning faith in the possibility of it becoming reality."[3] As we have already seen in the *Out of the Unknown* episode "No Place Like Earth," space flight is incidental, and the absence of "hard science" in it after the recent fly-by from the Mariner probe had upset critics of the episode.

Wyndham's interest in *Day of the Triffids* is less about exploring utopian frontiers in outer space than about the enjoyment of the possibilities of an eschatological fantasy, which suits its scientific romanticism. Fred Polak points out that there are two types of futuristic imagery in science fiction: utopian and eschatological, the second of which he discusses in his chapter "Devastation of the Image of the Future."[4] Eschatological thought had belonged mostly in the realm of theology—one thinks of the Book of Revelations—and in Mary Shelley's *The Last Man* (1826), about a plague that wipes out humankind. Wyndham's novel about the carnivorous Triffids discards the theological contemplation of "last things" to offer a secular acknowledgment of humankind's place in the universe. A contemporary of his would offer this justification for such fantasies:

> Work of this kind gives expression to thoughts and emotions which I think it good that we should entertain. It is sobering and cathartic to remember, now and then, our collective smallness, our apparent isolation, the apparent indifference of nature, the slow biological, geological and astronomical processes which may, in the long run, make many of our hopes (possibly some of our fears) ridiculous.[5]

It is this embedding of an eschatological fantasy within British scientific romance that produces the apocalyptic imagery of *Day of the Triffids*. At the same time, Wyndham's literary style is urbane and polished. It is easy to read and well-suited to the mass market of book clubs and paperbacks, which were burgeoning in the postwar period. The writing in the novel is a plain, semijournalistic prose that Wyndham learned from H. G. Wells. In turn, Wells had learned his trade in writing within the new journalism of the late nineteenth century that had been appearing in periodicals such as the *Fortnightly Review* and *Pall Mall Gazette*.

Like a tale recounted by Wells, *Day of the Triffids* is told to us by an English observer: the protagonist, Bill Masen, has the ability to see clearly the world as it is. It is an ability that allows the reader not to uncritically identify with the narrator, but rather to understand and observe the world with greater attention. Wyndham offers a clue as to his intentions when his hero Masen observes the dying city of London:

> My father once told me that before Hitler's war he used to go around London with his eyes more widely open than ever before, seeing the beauties of buildings that he had never noticed before—and saying goodbye to them. And now I had a similar feeling. But this was something worse. Much more than anyone could have hoped for had survived that war—but this was an enemy they would not survive. It was not wanton smashing and wilful burning that they waited for this time: it was simply the long, slow inevitable course of decay and collapse.[6]

There is plenty of sensationalism in *Triffids*, but it is underpinned by careful detail conveyed with a dispassionate intent.

> And so I came to Westminster. The deadness, the finish of it all, was italicised there. The usual scatter of abandoned vehicles lay about the streets. Very few people were in sight. I saw only three were moving. Two were tapping their way down to the gutters of Whitehall, the third was in Parliament Square. He was sitting close to Lincoln's statue, and clutching to him his dearest posses-

> sion—a side of bacon from which he was hacking a ragged slice with a blunt knife.[7]

In both passages, Wyndham's manner is almost airily detached, much as he was in real life. Many have described Wyndham to have been a diffident man who had inhabited the rooms of the Penn Club in London, which became his home for several years. In the novel, the luridness of the American pulps, which Wyndham had contributed to, has been discarded and substituted for a more refined style that promised veracity and suited the postwar period in England. The novel, like the 1981 television adaptation, trains the reader/viewer to accept the veracity of the account of the Triffids and the death of civilization. Brian Aldiss explains that the classic ground plan for a science fiction novel is veracity, capacity, and universality, and these structural characteristics can be found in both the novel and later adaptation.

In chapter one of the novel, there is a prevailing ambivalence about scientific discovery, which, due to the invention of the atomic bomb, had already produced the means of extinction. Wyndham makes clear that the nascent space race is the final stage of the opening of new horizons on Earth but, unlike the older conquest of frontiers, the satellites are a sign of technological progress but not necessarily social progress.

> It was by no means pleasant to realise that there was an unknown number of menaces up there over your head, quietly circling and circling until someone should arrange for them to drop—and that there was nothing to be done about them. Still, life has to go on—and novelty is a wonderfully short-lived thing. . . . From time to time there would be a panicky flare-up of expostulation when reports circulated that as well as satellites with atomic heads there were others with such things as crop diseases, cattle diseases, radioactive dusts, viruses, and infections not only of familiar kinds, but brand-new sorts recently thought up in laboratories, all floating around up there.[8]

What follows is an account of how new inventions in bioengineering will lead to disaster. The story is structured from the point of view of the protagonist, Bill Masen, a biologist, who by working on a Triffid farm has had the opportunity to study them. His realization that the Triffids possess the rudiments of intelligence suggests the possibility that humankind may be supplanted by another species for dominance of the Earth. The Triffids are a new species, whose origins are obscure although the suspicion is that they have been bioengineered in secret experiments in a totalitarian country (USSR). Bizarrely, they can move using a tripartite arrangement of roots, hence the name Triffids. On an overcrowded planet, which is desperate to feed the burgeoning human population, increasing at an exponential rate, it is discovered that the Triffids can be cultivated worldwide to produce valuable food oil, which is superior to existing fish and vegetable oils. Masen's discovery is that the Triffids often communicate by a rhythmic tapping of their roots, and he becomes wary of the distinctive sounds they can make.

THE DAY OF THE TRIFFIDS AS A "CLASSIC" TEXT

The story begins with a description of how nearly all the population of Earth go blind during a meteor storm, although the first line of the novel locates the reader in the quotidian: "When a day that you happen to know is Wednesday starts off by sounding like Sunday, there is something seriously wrong somewhere." Sunday was a day that would have loomed large in the British imagination when strict British shop and licensing laws made it a day of boredom and inactivity before starting the working week on Monday. Its perilous quietude was to be satirized a few years later by the comedian Tony Hancock.[9] "I *knew* things were awry," explains Masen. The location of the story in central London is another expressive feature in the novel that suggests a familiar setting while adding textual detail for the reader. The iconography of London would become important again in the two television adaptations of the novel, produced by the BBC in 1981 and 2009.

After some initial peregrinations, Masen and a young, attractive woman he meets, Josella, find a group of survivors at London University, but soon plague ravages the group and an attack on them leaves him free to relocate Josella. He is joined by a young girl, Susan, and eventually finds Josella in the county of Sussex. They settle down, but after several years find the struggle for survival more difficult. A helicopter lands and the pilot informs them of a successful colony on the Isle of Wight and asks them to join. They are reluctant to leave, but eventually a squad of soldiers arrives representing a despotic new government. Masen and the others disable the soldiers' vehicle and escape to the Isle of Wight. The anticlimactic ending—the mention of the Isle of Wight led Aldiss to declare the story to be a "cozy catastrophe"—was also criticized by Forrest J. Ackerman,[10] the editor and principal writer of the American magazine *Famous Monsters of Filmland*. The novel was judged by Groff Conklin, a leading science fiction anthologist, on its initial publication to be simply "a good run-of-the-mill affair."[11] Nevertheless, the novel would become a classic and enter the science fiction canon.

Wyndham's efforts at preventing his books being marketed as science fiction and cultivating an independent reputation recalls Graham Murdock's analysis of the role of the writer within television drama. For him, drawing on what had been done for the cinema, the ideology of authorship accorded a writer expressive autonomy, free of the commercial constraints of ratings. This approach helps to evaluate *Day of the Triffids* less as an example of the sci-fi genre and more as a cultural work that possessed enduring values rather than a topicality that becomes less relevant with age. The *Radio Times* would later describe it as a "20th-century classic."[12]

The marketing and promotion of the first television (1981) adaptation of *Day of the Triffids* positions the viewer on how they should interpret the to-be-aired series. The *Radio Times* was to remind the audience thirty years after its first publication, "[*The Day of the Triffids*] remains one of the most popular of all science fiction novels. With the start this week . . . it seems set for another three decades of excitement and terror."[13] The article continues by suggesting ways in which the viewer can think about the author: someone who was able to make the leap from "hack" work writing for the pulps to more legitimate work. The comparison within the *Radio Times* between Wyndham and Wells cements the claim of legitimacy, emphasizing what they have in common. Instead of excess, in the form of sex and violence expected in the pulps, the novel's main character is the unheroic and ordinary Bill Masen, much like the protagonists in Wells's early novels. In fact, Wyndham makes a sidelong reference in the novel to the lurid thrills expected in pulp sci-fi when he reveals that Josella is the author of a successful book called *Sex Is My Adventure*. Two large circulating libraries have banned it, "After that, its success had been assured, and the sales went rocketing up into the hundred thousands." Neither television adaptation (1981 or 2009) makes use of Wyndham's own attitude to salacious paperbacks and the pulps. Nevertheless, the subjectivity of Masen is retained in the adaptations.

HIERARCHY OF MEDIUMS: FILM ON TELEVISION

By the time of the 1981 adaptation, television production had been changing rapidly. One of the major changes had been the increasing production of drama on film shot on location outside of the electronic studio. We have seen that this was possible even as early as the 1960s in the episode *Stranger in the Family*, but what had been a tendency had become an established production culture that resisted the norms of television production within the studio.

During the 1970s, with the growing ascendancy of videotape, film became more of an old-fashioned, outdated, and messy medium. And yet, it refused to disappear. Film continued to be regarded as a medium suitable for the cinema, whereas videotape was the proper form for television. The motion-picture industry is far older than television and it was where film had acquired the status of a long-established art. Film was regarded as "art," whereas video was "craft." Video had not attained the connotations of the feature film industry. At the same time, this view that sees film as antithetical to television must be understood as one of the many complex ways that film was being used. Live production, videotape recording of live production, and the shot-on-film series had assumed complementary roles in the pattern of television scheduling during the 1960s. As videotape improved in quality, and it became possible to edit video, a new production medium also took form. Video came to be regarded as more than just a method of rebroadcasting live TV programs; it appealed to producers as a substitute for film itself. The key was in the editing process.

Since the 1970s, it was possible to be more creative by allowing for the postproduction process to be less pressured than editing on-line on two-inch quad tape. As it became easier to edit videotape, it allowed for professional commercial film editors rather than video tape recorder (VTR) editors from television to be made available. The fact that a helical VTR could be played frame-by-frame, in slow motion or fast motion, forward and reverse, was one of the reasons why film editors were able to adapt to videotape editing.[14] Off-line editing involved the transfer of material to low-band tape, so that editing decisions could be made in a low-cost situation. Later, the on-line edit of the editing of the two-inch master would be on quadruplex machines at the various videotape facility houses, supervised by television-grade videotape editors.

However, the digitization of the video image, which promised tape-quality that was closer to the high image resolution of 35 mm film was still to arrive and would take another two decades. Indeed, film continued to fulfill the needs of the program makers of the 1980s by producing a "high-quality" product that allowed the program maker to work unconstrained by the many technical parameters caused by shooting on video in a studio, thereby limiting their creative ability. By the end of the 1970s, out of a BBC production total of 12,000 hours of programming per year, 35 percent was live, 40 percent was produced on tape, and 25 percent was produced on film.[15] Already, by the early 1960s, the television camera had shrunk until it was no bulkier and no less maneuverable than a 35 mm film camera. The key difference between an electronic camera and a film camera was that a mass of cables had to be run from the television camera to the portable monitors and the Ampex equipment used to videotape the program. This tended to impose limits on how the electronic television camera might be used, especially on location. Moreover, with videotape (Sony's U-matic format from 1971) there was never any possibility that the picture quality could approach the richness of 35 mm film.[16]

By the mid 1970s the notion of "good television" was equated by Stuart Hall, the British cultural theorist, with "professional" and high technical standards.[17] For Hall, this would serve television's need to fulfill its commitment to mass entertainment without the distraction of a low-quality picture that would continue to suggest an invisible technical medium or relay to an audience.[18] In 1976, Brian Clark, a BBC producer, sat on the jury at the Royal Television Society to choose the best single play from a shortlist comprised of four—all on film—which represented, probably, at most, 10 percent of the "single play" drama output on British television. They were all filmed single plays, and the small percentage of British drama output that happened to be on film also happened to be those that the viewers liked best. For Peter Eckersley, also on the panel, this wasn't a coincidence. "Production standards have gone up enormously and one can't get away with things that don't look very good." For the panel, "The best single play year after year is on film, but not with the serials. Amongst the serials, all four of them were largely studio."[19] The studio had been the only affordable and viable method in the 1950s of recording the majority of drama until the introduction of 16 mm film, and by the end of the 1970s, drama production had moved decisively toward film. The fact that it was shot on film initiated a marked rise in production values—the look, feel, and atmosphere of a "realistic" drama shot on location—compared to a studio-based drama recorded in the 1960s.

From the 1960s, the technology of 35 mm and 16 mm film was adapted to television as that medium became the principal user of celluloid rather than the cinema. New film stocks meant that color and print density were made available to television producers. On British television, there had been a steady annual increase in the use of film since the 1970s. The amount of raw film stock used and actually transmitted by 1980, allowing for shooting ratios of between 7:1 and 20:1, would have made 2,150 average-length feature films.[20]

The decision to shoot a drama production on film and the evaluation of the cultural value of a shot-on-film program had become fully entwined by the time of the 1981 production of *The Day of the Triffids*. As a medium, television, which had been once characterized as one of relay and communication rather than an expressive medium in its own right, had gone on to establish a hierarchy of programs. Such hierarchies resisted the idea of flow or the dispersal of programs stripped of their individual characteristics. It had also been noticed by television producers that although immediacy was important to a nondramatic program, it was not necessary for drama. A precise sense of immediacy made very little difference in much of television drama, although topicality and realism continued to have wide-ranging appeal for television's drama audience.

In Britain, the increasing use of 16 mm film and the desire for greater "authenticity" in television drama had led to many practitioners identifying the studio with producing entertainment or routine programs such as soaps. At the same time, since the 1960s, the use of 16 mm had been professionalized. Jamie Sexton has pointed out that the original difficulty of using 16 mm for television drama was its association with amateurism.[21] Although increasingly professionalized by continued technical improvements, it was believed to be an amateur gauge because of its low image resolution. News and current affairs used 16mm film more frequently, but it continued to be avoided in the recording of drama because the technical quality of the image in drama was believed to be important due to its aesthetic requirements. It was because film production for the cinema had switched to television that filming techniques were not only adapted to television, but also were later used to devise television's own modalities by employing 16 mm film. Programs shot on 16 mm film were considered to be distinct from cinema films and made to occupy a different aesthetic space between the cinema film and the studio-bound "live" television drama. The identification of 16 mm film with social realism, as we have already seen, can be argued to have marginalized other forms of television, such as the glossy telefantasy series shot on 35 mm film that was exported.[22]

The BBC had adapted the *Triffids* for radio as a reading or as dramatization in 1953, 1957, 1968, 1971, 1973, and 1980. Each adaptation tended to be faithful to the original novel with few true additions. The 1968 adaptation begins differently from the novel, but such differences are about plotting rather than adding material that is noncanonical. By 1979, when the decision had been made to adapt it for television, it had become classic genre material. Producer David Maloney was allowed to make the series once an ambitious cofinance deal had been agreed between the BBC, the Australian Broadcasting Corporation, and the American cable network RCTV, which would cofinance with the BBC the following year another classic serial, *The Woman in White* by Willkie Collins. David Livingstone was the writer, and he made few changes to Wyndham's novel. It was to be a faithful adaptation, reemphasizing the cultural value of the story and its iconic status. The decision was however made to update the novel from the 1950s to the present (1981). Some minor changes were made: Josella Payton was originally a wealthy author of a novel entitled *Sex Is My Adventure*. Instead, she became a modern career woman. Though she was still wealthy, the television series was intended for a family audience. More significantly, the adaptation, like the book, preferred a low-key style that was naturalistic in its observation of realistic detail and possessed a keen sense of the everyday. The protagonist, Bill Masen, played by John Duttine, became the Everyman of the story, equipped with a Yorkshire accent and lacking the usual screen magnetism reserved for a leading man. Masen is a diffident man, although it emerges that he has a strong sense of social responsibility and is practical enough to possess leadership qualities.

Livingstone was to deliver an adaptation that matched the observational realism of the book and its sensibility of private individual drama played against an epic scenario. Unlike American sci-fi series at the time, it lacked special effects and was more a meditation of post-catastrophe Britain. The lack of special effects meant that the eponymous Triffids are kept in the background and the action is centered on the struggle for survival after the failure of civilization and a reversion to savagery. The adaptation made in 2009 would reverse this order. In 1981, the reaction of the critics was to be mildly critical of the adaptation. The Triffids were thought to be improbable, suggesting a failure of the verisimilitude used within the series. For example, Herbert Kretzmer, the television critic of the *Daily Mail*, claimed the plants "looked no more menacing than a bunch of outsized orchids left over from the *Muppet Show*."[23] Such reaction to the show was evidence of the changes in television that demanded higher production values in a series that aspired to be an example of "quality drama."

By the mid-1980s, there was a second tendency for "quality" drama to not only be shot on film, but also to align itself with the cinema. A quality serial embodied notions of "art" and "significance," such as the period drama *The Jewel in the Crown* (1984), with the type of production values that produced the "look" and "feel" that audiences by that time would have expected. John Walker, cameraman on the BBC's detective series *Shoestring* (1979) was to say "The film camera goes where the viewer will be and captures the atmosphere. It is more real on film."[24] Derek Granger, the producer of *Brideshead Revisited* (1981) agreed: "[Film] provides us with a depth of feeling, the realness that we are seeking. It has a feather-like quality that video doesn't have. . . . Aesthetically, video is a coarse medium, very suitable for some types of production but totally inappropriate for this one."[25] In these comments there was an implicit hierarchy of media on television and an argument for the ascendancy of film.

The need for coproductions was driven by the rise in production values demanded by expensive drama. In 1952, seven hours of Bernard Shaw's *Back to Methuselah* had had a BBC budget of £3,400, the equivalent of about £80,000.[26] At the end of the 1970s, as we have seen with the deployment of film for up-market productions, a top range of productions were being allocated proportionally larger resources than ever before. By 1981, the BBC assumed that each ten minutes of 16 mm film would cost £100. Consequently, the cost of film alone for *Triffids* at a shooting ratio as low as 10:1 (50:1 was more common in expensive drama) would be about £180,000. Aesthetically, prestigious television increasingly aspired to look like cinematic film, which suggests the beginning of a new alignment or possible convergence between cinema and television.

> The prestige one-off drama series *Armchair Theatre* cost only three times as much as the soap *Coronation Street* at the turn of the 1960s. A prestige drama series episode currently costs five or six times the cost of *Coronation Street* [2005], and a single drama would be made more like a cinema film with a considerably larger budget. So it is immediately clear that the range of available budgets for producers has substantially increased.[27]

Unfortunately, the range of locations, the professional camera standards of image density and color values, and the ease of postproduction meant that since the 1960s there had been a relentless rise in the real costs of producing television drama. A consequence of the increase in budgets, as well as costs, was a developing tendency for more coproduction projects, which also saw a new spirit, within as well as outside of television, of internationalism on the threshold of the Direct Satellite Broadcasting (DBS) age. Each development meant that British, American, and European companies were combining different resources to create lavishly packaged, large-budget entertainment that could command a continued, loyal viewing by the audience.

The rise in production costs had also affected less prestigious projects, including *Triffids.* The new-technology markets of cable television and domestic video were beginning to attract an audience interested in obtaining quality drama that could be watched again and again, including adaptations. The long form had not been particularly profitable, unlike the series, because repeats on broadcast television tended to have below-average ratings. Moreover, the styles and subjects of the long-form drama had to remain distinct from such serial forms as soap operas. Whereas lush locales and dazzling costumes had once been confined to the long form as miniseries, by 1981 such production values had become visible in other weekly series. Perceived program value was linked to the idea that what could be released on video would have enduring values, and a "quality" long-form drama was ideal for repeated viewings on video as well as cable. However, *Triffids* was produced as six, twenty-six-minute episodes, which had been planned so as to not disrupt the schedule in the same way as a lavish miniseries. While *Triffids* is a "special" adapted from a well-known work, it was shot more as a program to fit television's schedule, without disrupting the flow of already-established programs at their regular times. Such a series restricted to a preexisting schedule would be left behind by the development in the United States of such glossy series as *St. Elsewhere*, *L. A. Law*, and *Crime Story*, which could maximize profit and minimize risk by borrowing from other forms, as well as breaking barriers to cover more "adult" subjects.

For these reasons, the difference in production values was not as marked yet as it would be in the second (2009) adaptation. Instead, the 1981 adaptation of the *Triffids* sought an indigenous form, an observational naturalism shot on 16 mm, which could be used to represent a "gritty realism." The origins of gritty realism in British television drama as a distinctive aesthetic are beyond the scope of this book, but the desire to present the story as an eschatological fantasy would mean that the program makers would be naturally influenced by it.[28]

GRITTY REALISM

The visual style of *Triffids* owes a debt to the social realist drama that had been shot on film since the 1960s at both the BBC and ITV. The style is not to create so much a greater realism as a *gritty* realism. The result is a realistic "serious" drama that deals with psychological depth, while continuing to contain enough action to be watchable. In many ways, television drama is the antidote to the spectacle of the cinema and depends on content and character rather than plot and action. For example, the opening of episode one of the *Triffids* is restrained, without the benefit of a large set-piece spectacle to encourage the viewer to continue watching. Instead, we see Bill Masen in a hospital bed and, using the device of talking into a portable tape machine, he narrates some expository facts. In many ways, the soliloquy echoes a radio adaptation, using voice rather than image to convey its meaning. No cinematic master shot is used to establish Masen's whereabouts; it is left to the audience to imagine the off-screen space of the hospital he finds himself in. The centrality of the performance establishes the importance of character and lends depth to the story. Masen's dictation to the tape machine cuts to exposition about the origins of the Triffids and how they became a valuable commodity to be farmed for their oil. The dense, slow-paced exposition continues with how it was discovered that the Triffids are carnivores before moving on to the mass blindness caused by the gigantic display of meteors.

The adaptation of an indigenous style in *Triffids* can be compared to the making, a few years earlier (1979), of the final episode of *Quatermass*, based on a new script by Nigel Kneale. The idea of bringing back Professor Quatermass for a fourth adventure dates back to when Irene Shubik had asked Kneale to contribute a new story for the first season of *Out of the Unknown*. The offer had fallen into abeyance when in 1972 Kneale had been asked by the BBC to write a new four-part *Quatermass* serial. The story was set in the near future amid urban decay and anarchy; an extrapolation of the "violent society" and turmoil of the early 1970s. Kneale's theme of the "drop-out" society had been used in an earlier science fiction television drama, *The Year of the Sex Olympics*, which he had penned in 1968 and had been shown on BBC-2 as part of *Theatre 625*. Similarly, it shares a concern for the outcome of the counterculture of the 1960s.

The production experience for the new series of *Quatermass* had been an unhappy one according to Ted Childs, who had produced it for ITV. Twenty-five years after the transmission of the original *Quatermass*, the BBC had chosen not to make it because of cost. Verity Lambert, who commissioned it for ITV, explained:

> *Quatermass* came to us in the shape of four scripts which had been commissioned by the BBC and which had been turned down by them as being too expensive for them to make. It was very expensive—I think it was the most expensive thing we had attempted at Euston Films at that point. And we felt that the only way we could justify the expense was to make sure that we could re-edit it into a film which could possibly have theatrical release as well as have a four-parter [for TV].[29]

Since 1964, one of the most important stimuli to TV filmed production in Britain had been the Exchequer Levy on ITV contractors' profits. Under the Television Act of 1964, the first version of the levy on advertising revenue was introduced. However, certain expenses could be set against the Levy and one of these had been the cost of programs whose earnings from overseas sales were themselves from the Levy, as a contribution to the national export drive. This meant that at times up to two-thirds of the cost of a filmed drama or "television feature" could be deducted from the companies' Levy payment.

> It would seem that this, more than anything else, was the reason for the rash of film production subsidiaries being created by the major ITV companies during the 70s. ITC and Thames' Euston Films have been the most active and successful, though both producing mainly series-drama; while Southern Television's Southern Pictures, ATV's Black Lion, and Granada and STV off-shoots have all experimented with full-scale British TV movies, destined for television only in Britain, yet available for theatrical release, as well as television, abroad.[30]

A gritty style was appropriate for the new *Quatermass*, but a decision had been made to produce it with high production values and adopt a more cinematic style to enable it to be exported overseas. Unlike the original *Quatermass* series or the BBC's *Day of the Triffids*, the series was allocated £1 million (a typical budget would have been a tenth of that) and the entire series was shot on 35 mm Panavision stock. However, unlike a large-budget film, which could be shot on a sound stage at Pinewood or Shepperton studios, the new series of *Quatermass* would rely on the objectivity of the camera and deploy a stylized, fictional realism. Such objectivity was to be achieved by shooting on location, although the British government's Department of the Environment had refused permission to shoot the climactic scenes at the ancient monument of Stonehenge.

Ted Childs identified the lack of success of *Quatermass* as due to its being inappropriate to a medium whose reception was in the home. On this occasion, its challenge to television's essential mundanity or to its day-to-day regularity was to prove its undoing:

> I think there were high production values in it and I think that the performances were good but the primary problems with it were a) it was perhaps too depressing a story for a popular television audience and b) the punters were used to a fairly high standard of technical representation from American television—you know you had had *Star Trek* and then there had been movies like *Star Wars* and *Close Encounters of the Third Kind*. And we couldn't afford that.[31]

Although television can often deny its origins in popular culture by offering event programs that lift it out of the flow, such shows like *Quatermass* require a budget and production resources that complement the possibilities of large-scale narrative. Such large-scale narrative evokes the scale of the best-selling international novel, but *The Day of the Triffids* was, in fact, ideal for television adaptation for a domestic audience at home because of its relatively low-key, observational based veracity.

SPECIAL EFFECTS AND SCIENCE FICTION: *THE TRIPODS*

> It might be asked, do cinematic codes of visibility and developments in special effects and in sound and color technologies connect with thematic transformations of the genre? How do narrative and spectacle, in their transformation and their interrelation, intersect with the cultural instrumentalities of science fiction cinema?[32]

The possibility that the mode of production can affect genre has already been explored, but if the *Triffids* BBC adaptation in 1981 had been produced using a bleak gritty realism, then the development of digital technology transformed the visibility of science fiction on television. The digital manipulation of images and sounds would also affect the production process. As soon as video is digitized, images can be manipulated, for example, overlapped with other images. Sound can also be cleaned of unwanted background noise or a technical problem, and edited with the click of a mouse. A major impact on production is that shots become framed and composed according to the special effect to be added. Some directors will add a video camera or "tap" to the film camera to see scenes and experiment with trial edits before the film is processed. To this extent, the 1980s marked the beginning of the production phase merging with the postproduction phase.

The novel *The Tripods* was written by John Christopher, a pseudonym for Sam Youd. He had been writing science fiction since 1951, beginning with short stories, much as John Wyndham had twenty years earlier. His first novel, *Year of the Comet*, was published in 1955. His second novel, *The Death of Grass* (1956), was to mine the same rich vein of post-apocalypse possibilities that Wyndham had used in the *Triffids*. Once again, *Death of Grass* is about the effect of science on Earth's biology, this time, a virus kills off all forms of grass. The novel was later adapted into a film produced by MGM. The struggle of a British family to survive in a world where food and sanctuary seem to have disappeared underlines the similarity between Christopher and Wyndham in their approach to science fiction during this period.

In 1966, Christopher started writing science fiction for adolescents. *The Tripods* trilogy (1967–1968) was a variation on the postapocalyptic theme the author had been using. The story is about humanity enslaved by the

Tripods, giant tri-pedal walking machines that are piloted by unseen aliens, identified only as the "Masters." Civilization has reverted to a pastoral life under the watchful presence of the Tripods, with few habitations larger than villages. The population of the Earth is controlled by implanted "Caps," which are installed in each person at the age of fourteen, suppressing any independent thought. The novels were published as *The White Mountains* (1967), *The City of Gold and Lead* (1968), and *The Pool of Fire* (1968). In the books, the adventures of the three main characters, Will, Henry, and Beanpole, take place as they journey from England to France in search of a method of battling and finally defeating the Tripods.

The BBC decided to produce *The Tripods* in 1982, and season one was broadcast in 1984 as thirteen half-hour episodes. It was scripted by Alick Rowe and produced by Richard Bates. Seasons one and two both attracted substantial coproduction finance. The series, which had been touted as a successor to *Dr. Who*, was a coproduction with the Australian Seven Network and used the services of international distributor Freemantle International. Freemantle was to sell *Tripods* to over forty-one markets in the United States, as well as to France, Sweden, Denmark, Switzerland, and Belgium. As we have seen, the purpose of coproductions allows for increased foreign sales in order to maximize profit and minimize risk. Paul Talbot, then president of Freemantle International in New York, explained: "The series is selling moderately well in continental Europe and has achieved a 45 percent penetration in syndication in America. . . . I would eventually hope to achieve a penetration of 70 percent in syndication."[33]

The series became one of the BBC's top-rated children's programs in 1984, peaking at an audience of nine million. On this success, the second series of *The Tripods* was budgeted at more than £1 million, which hitherto had represented the BBC's biggest investment in science fiction. However, as was usual for television, the series had been conceived less as science fiction and more as an imaginative adventure story.

Earlier television production had suffered from various technical limitations, but with improvements in video production technology and technique, videotape afforded television producers the flexibility of film without the cost, while still retaining the aura of liveness. However, in spite of *Tripods* being recorded on videotape and the formal expectations of what television is, the story events are presented in a particular way in order to thrill. Viewer expectations include chase sequences, various types of combat, and a greater pace in the editing, although in *Tripods* the latter is not always apparent. As an adventure story, *Tripods* allows for a journey or quest to be made by the protagonists, and provides a stage for several types of action.

The method of presenting narrative in television touches on the specificity of science fiction itself and how it seeks to affirm its own distinctive status. However, the recourse to action as a form of spectacle suggests less of a distinction between science fiction and adventure. Equally, in television, as the science fiction story becomes less concerned with the staples of genre, the medium's phenomenology and the erosion of generic boundaries complicates the understanding of how genre operates. In fact, rather than think of adventure as a genre, Brian Taves discusses how it can be understood as a mode:

> The usual definitions of adventure stress elements of the unusual, overcoming obstacles with narrow escapes, and vanquishing villains. In this sense, adventure becomes linked with action, a word attached to any film with a greater emphasis on action than emotion. Indeed, action is a more appropriate word than adventure to describe the style of storytelling that runs through many genres, a male-oriented approach dependent on physical movement, violence, and suspense, with often perfunctory motivation and romance. Action tends to shift sentiment, character, dialogue, and family to the background.[34]

In this way, adventure can be multigeneric. Television, which often relies on "flow," itself based on scheduling, does not make an appeal based on a single program and the staples of a single genre. Indeed, relying on this logic, television's main contribution is less about formal expectations. Rather, its main contribution is the new relationship it creates between imaginary worlds and the audience. As we have already seen, television's imaginary worlds can attain long-term, ongoing integration into the lives of its audiences. Nevertheless, television continues to supply originating material.

Richard Bates had worked on the idea of an adaptation of *Tripods* for fifteen years, and two years had been needed to plan the necessary special effects. The BBC had agreed to the production if the series could be recorded on videotape rather than shot on film, in the same way that similar shows such as *Doctor Who* were taped, unlike the more expensive option of shooting on film. Yet Bates believed that the decision to record on tape was one of the keys to a successful adaptation for television. Originally, optics or special effects had been more difficult to do on tape than on film, but technical innovation had led to greater ease with adding effects. He was able to say about the decision to record on tape: "It was exciting to work on the frontiers of known video technology. The production team was using modern computer technology, computer-assisted design and a range of new technical equipment including the Quantel Paintbox."[35] Such equipment was equipped with an artist's stylus and a graphics pad rather than a keyboard in order to suggest an intuitive and emotive interface with the technology. The new digital devices were not simply to be a faster, more efficient method of working, but a promise that the equipment could encourage greater creativity.

The Tripods marks a moment (but not the only one) when the use of digital effects to manipulate images had overtaken the desire within television to maintain a sense of the transparency of televisual imagery. Rather than television's privileged myths of liveness and transparency, the electronic postproduction environment consisting of styluses, track balls, and digitizers, allowed for an array of effects. Quantel, a manufacturer since 1973 of digital production equipment for broadcast television, advertised the capabilities of its Paintbox as: "Every type of artist's medium is provided by the menu and stylus: paint, chalk, airbrush, wash or shade. Pictures can also be magnified, blurred, crisped, or smoothed to achieve just the desired effect. Stamp, smear, smudge."[36] Invented in 1981, during the production of *Tripods* there were only three Paintboxes in Britain.

Besides digital postproduction, two models were used for the Tripods, the metal monsters that rule humankind after its conquest. There was a two-foot model and a larger one, which at full size would have been ten feet but did not possess legs. Three full-size half legs that were thirty feet high were used on location and shots of the models were matted in during postproduction to complete the Tripods. In this way, composite visual effects could be created to appear as solid devices on the screen. For Bates, such expensive techniques were possible because the BBC had the resources and expertise for its use. The major advantage of the Quantel Paintbox was its ability to make changes to the composite images automatically, unlike the creation of individual mattes on film, which each time required a separate effect. The postproduction effects for *Tripods* were completed at the BBC's Video Effects Workshop, forming a sophisticated electronic control room.

The creation of the Tripods, using a mix of digital effects and animatronics, helped to establish a world of illusionistic realism and generic verisimilitude. Digital effects would be used to create transparent layering, a tactic that offers a mode of immediacy and transparency rather than a mediated image. In this way, in spite of the use of digital and other sophisticated effects, televisual style continues to be media specific. The television adaptation of *Tripods*, because it was relatively low-key, retained an intimacy with its main characters in spite of its willingness to embrace some aspects of a cinematic epic due to its greater use of spectacle. However, instead of cinematic spectacle, there are clear ontological differences. We are presented with a videotaped program, and the length of the episodes (thirty minutes) engages the audience in ways more suitable to a medium situated in the home.

Other decisions were made to alter the television series from the book. For example, in order to not overuse the Tripods, the Black Guard was introduced. It was expensive to film the Tripods, and the Black Guard was conceived as a substitute menace. Nevertheless, the budget for the series was sufficient to include other new techniques such as animatronics and the building of the Masters' City, which up until that time was the largest model the BBC had built, covering twelve thousand square feet, filling an entire studio at Ealing. Paintbox was used to combine a shot of a walkway measuring a hundred yards on top of images of the model city so as to create a sense of scale.

During the mid 1980s, the challenge to the BBC was not only from technical innovation, which threatened to alter the form of television drama, but also political changes to the structure of the institution. After winning a second election victory in 1983, the Conservative government of Margaret Thatcher sought to weaken the principle of public service broadcasting based solely on funding from what was claimed to be the outdated license fee. Instead, the government sought to explore alternative methods by which the corporation might be funded, in particular the introduction of advertising. A committee of inquiry led by Sir Alan Peacock was, according to Peter Goodwin, the first report in the history of British broadcasting to see the market rather than public service as the underlying principle of broadcasting. *Tripods*, as an expensive, special effects–laden series, catering to an audience not regarded as natural BBC territory, and therefore lacking the support and favor of senior BBC executives, discovered that it was vulnerable. A memo from Richard Bates expressed the concerns held by the makers of the series:

> Whilst the series [Tripods] has struggled against strong competition to reach a large audience, it has not been a disaster. Neither has it been expensive; Series One and Two having attracted substantial co-production finance. Nevertheless a change in Program Controller and Head of Department has prompted a very different attitude to the production and I must argue that this is both demoralising for BBC staff and professionally immoral on the part of the management.
>
> I am genuinely sorry that my Head of Department does not like the series but I cannot mutely accept that he has the right to sweep away programs encouraged and developed by his predecessors. I recently informed him that our co-producers are willing to meet all External Costs for Series Three but have received no response from him to this exciting news.[37]

The reply from Jonathan Powell, the Head of Series/Serials, revealed the principles by which the BBC believed it would direct future decisions about production:

> Quite frankly in our view the program does not justify its continuation either in terms of its competitive strength or its editorial merits or both.
>
> It is a well and interestingly made piece but the second series hasn't substantially improved on the shortcomings of the first in the areas of character realisation, acting and narrative. . . .
>
> I cannot be bound by decisions taken in principle by my predecessor, particularly when faced by long commitments which do not live up to their promise.[38]

While criticism of *Tripods* can be argued to be justified, its fate demonstrates how the BBC sought to justify its continuing method of public service funding. Competition from a multichannel digital broadcasting landscape was to culminate in an attempt to no longer fund drama of insufficient "quality." Accordingly, if *Tripods* required the viewer to remember many of the details it created, its long form over many episodes serving this narrative aspect especially well as an instance of an entirely imaginary world, it failed to achieve the high degree of immersion that the long form of drama was meant to deliver.

FEAR FACTOR?

With memories of the 1980s series, the BBC was to remake *The Day of the Triffids* in 2009 as part of its Christmas season. It was the third major adaptation of the *Triffids*, following the 1962 film, the 1968 BBC radio program, and 1981 television series. The BBC was to coproduce the new adaptation with European producer/distributor Power and Canadian production company Prodigy Pictures. The projected scale of the new show and the method of coproduction was a sign of the increased need to appeal to an international television marketplace. In this way, *The Day of the Triffids* becomes a self-consciously transnational drama, as well as less a self-conscious piece of science fiction drama.

> Though national borders have not been entirely eroded . . . it makes little sense in the context of a global market-place to consider the political economy of any specific nation in isolation. Today's production and distribution is vertically and horizontally integrated within the "culture industries" on a global scale. Little more than a decade ago, however, systems of finance, regulation and production were organised nationally, even though there was evidently some exchange of product and limited co-production. Generally speaking, apart perhaps from in the USA, dissemination was conceived primarily on a basis of national cultures and consumption. Exports were desirable and brought additional profits but they were not typically inherent in the conception of projects. Building on significant changes in the 1990s, TV3 [the third age of television] is characterised, by contrast, by national industry deregulation which effectively facilitates bigger companies with transnational vision and reach.[39]

The production was shot over nine weeks during the early part of the year, in and around London. Patrick Harbinson's script was to maintain the original idea that in the near future the Triffids are farmed to produce natural oil, Triffoil, after genetic modification. The oil is used to stop global warming. The budget was large, £15 million, but not by the standards of a film laden with special effects. Yet the budget was to pay for an expensive cast that included Dougray Scott, who had previously appeared in the Hollywood films *Enigma* (2001) and *Mission: Impossible* (2000). It also included Vanessa Redgrave as Christine Durrant and Eddie Izzard as the villain, Torrence, as well as Joely Richardson playing Jo Playton.

The role of Bill Masen was played by Scott, who was expected to act in front of a green screen. The Triffids would be computer generated and added in postproduction to appear that they were with him. Since the days of Paintbox, digital technology had become capable of producing a sophisticated rendering of professional key and matte effects. Rather than rely on painting constructed facades or going on location, it had become possible for directors to shoot their actors on relatively small blue or green sets. It was then possible to digitally composite the background into the images. On the 2009 adaptation of *Triffids*, one image could consist of several layers of composited video.

Such advances in special effects technology altered the type and range of spectacle that was possible in earlier television science fiction. The producer of *Triffids*, Stephen Smallwood, explained that more than anything the Triffids had to be plausible: "The brief to the designers was to come up with a Triffid that is a completely believable plant, not something that has come from Mars. . . . If you look at the previous attempts at Triffid depiction, the stinging is plausible, the movement is not. We've created a plant that has some validity."[40] The promise of plausibility, linked to generic verisimilitude, can be understood here as the flaunting of digital production technologies in order to realize a more realistic representation of the monstrous. The director of photography, Ben Smithard, was to remark:

> The earlier versions didn't have any ambitions visually. The producers kept it very small scale. However much I love television, I don't love bad television. I really hate it. That was why I didn't want to make it big; I wanted to make it huge. We tried to make it look like a big American movie; something Spielberg would do, like *War of the Worlds*. We wanted to grab the audience's attention and throttle them with strong imagery.[41]

Instead of adapting *Day of the Triffids* according to the methods and style of television drama that had been codified earlier, Smithard explained,

> Television's always been shot in 16mm, which is a very good medium, but not a big medium . . . I want to put everything on the screen; the same as when I shot *Cranford* [a BBC period drama]. I didn't want to shoot Judi Dench on 16mm; I wanted to see every line on her face, for better or worse. The same with this. I want to get all the production design, costumes and make-up on the screen. If you did project this, it would be just as good as any movie. That's the idea.[42]

The director, Nick Copus, was to add that "while there is an interesting message at the heart of it, the new *Triffids* is a popcorn movie. Whereas the eighties was bleak and minimalist, we've gone the other way."[43] If the former adaptation had thrived on its minimalist appeal and purely functional characters, the attempt was made in the new version to upscale the drama, and this can be seen in the opening credits, which come after an initial, extended teaser.

The 2009 adaptation was a response to the challenge of representing the bizarre on television using the possibilities of digital technology. For John Caldwell, television in the 1980s and 1990s was able to use innovations in audiovisual media, which he argues was the development of a new "aesthetic and presentational task."

> Television has come to flaunt and display style. Programs battle for identifiable style markers and distinct looks in order to gain audience share within the competitive broadcast flow. . . . The stylistic emphasis that emerged during this period resulted from a number of interrelated tendencies and changes: in the industry's mode of production, in programming practice, in the audience and its expectations and in an economic crisis in [US] network television.[44]

Caldwell concludes that the term *televisuality* describes a stylizing performance because it can offer more visual and aural options than previous modes of production within television. Although he does not believe that developments in technology automatically cause formal changes and image and sound are not by-products of technical evolution, it should be remembered that the tremendous advantage of digital technology is its ability to generate options for shot sequences, effects, and sound at the postproduction stage. The ability to manipulate pictures and sound does not substitute for careful setup and shooting, but digital postproduction can darken images, brighten them, enhance color, and crucially overlap an image with other images that would have taken weeks to complete in a film lab.

The opening credits of *The Day of the Triffids* demonstrate the ambition of the show not only visually, but also aurally. The orchestral music offers a fuller sound and an intensification of the entertainment value of the show. This undercuts the proposition given by John Ellis about the specific way television has been watched:

> The broadcast TV viewer is not engaged by TV representations to any great degree: broadcast TV has not so far produced a group of telephiles to match the cinephiles who have seen everything and know the least inconsequential detail about the most obscure actors and directors. Broadcast TV does not habitually offer any great incentives to start watching TV; instead, it offers them to people already watching TV. . . . Where cinema solicits its spectators outside the screening of a specific film by the construction of a narrative image, TV draws the interest of its viewers through its own operations of broadcasting. The viewer is cast as someone who has the TV switched on, but

> is giving it very little attention: a casual viewer relaxing at home in the midst of the family group.[45]

Instead, as the medium of television has changed, the drama of the third age of television (TV3) engages its audience in ways that are different from earlier televisual forms. By offering a more complex style, the viewer's experience of television becomes more valuable as he or she is asked to work harder to negotiate meaning from the text.

For example, the design concept for the giant Triffids was left to Joel Collins, the producer designer. This involved visits to the Eden natural environment project in the British county of Cornwall and a careful viewing of the BBC's *The Private Life of Plants.*[46] A study of the growth, movement, reproduction, and survival of plants, *The Private Life of Plants* was the second of David Attenborough's specialized surveys following his major documentary trilogy *Life on Earth* (1979), *Life in the Balance* (1984), and *Trials of Life* (1990). *Life of Plants*, a natural history epic, combined objectivity with the bizarre beauty of plants. The materialist objectivity of documentaries became collapsed in the fictional drama of *Day of the Triffids*. It deployed the artifice of CGI (computer generated imagery) to form a seamless blend of images in order to create a coherent diegetic world, allowing for the Triffids to be both plausible and physically real. In this way, the ideation of the Triffids adheres to both the dual standards of perceptual and cognitive realism.

Seamless images also have the ability of offering highly spectacular digital effects. The imagery of the Triffids was to be highly observable and allowed to dominate the narrative when required. This was unlike older adaptations when the appearance of the Triffids had to be more restrained or the monstrous in science fiction was to be altogether avoided. Irene Shubik had argued that the bug-eyed monster (BEM) was unsuitable for science fiction appealing to adults, which instead should offer original and philosophical ideas. The issue becomes whether CGI is being used as a substitute for a story that makes particular demands on the audience and if it is being asked to work harder. Nevertheless, the use of CGI was suitable for an extraordinary tale in which impossible possibilities become visible. In fact, the use of CGI continues to create a perception of verisimilitude in extraordinary circumstances. If there is a resistance to breaking the illusion of a realistic world on the verge of destruction, CGI is used as an extension of the real-world environment to increase perceptual realism. Albert La Valley remarks that this use of CGI "show us things to be untrue, but show them to us with such conviction that we believe them to be real."[47]

The production team concluded that because the Triffids are capable of perambulation, they would be sinuous. As Smallwood explained, "being entangled in the massive roots is one of the beasts' principle ways of catching people."[48] The Triffids would be thin and upright and about twelve feet tall. Finally, the top of the Triffids was developed with a wisteria-type neck supporting a "hooded cowl." "It's a bit like a hoodie," explained Smallwood, "out of which comes a scary stinger." Unfortunately, a review of the show by the British newspaper the *Guardian*, remarked that the killer plants were more like "asbo cacti." The review commented that "rising from the centre of the plant, is a kind of red hoodie—possibly playing, like a *Daily Mail* editorial, to our fear of feral youths. The Day of the Asbo Cacti. . . . They don't frighten me."[49]

One problem encountered by the audience and the use of CGI is, as Joel Black observes, "the graphical imperative in present day cinema whereby everything must be shown and as little as possible left to the viewer's imagination."[50] Similarly, the danger is that television drama can because of CGI lead to reliance upon spectacle as a substitute for character or story-driven action or else excessive action and effects sequences. There is also the danger that too many special effects stop the drama feeling like television. There exists an unresolved tension between *Triffids* appearing as a television drama and the foregrounding of its visual effects. According to Caldwell, this tension has been resolved by the decision to adhere to a *smart style*. For him, arguments from *American Cinematographer* offer an alternative to the aversion of style on television and lack of formal innovation, which he believes should be understood as a legacy of television's original relay function. Such a smart style offers self-consciousness about television that is not unmotivated but suggests that production personnel in television are more stylistically and theoretically inclined than has been the case historically. Television, conceived in these terms, exists as a medium that, despite using similar production techniques as cinema, avoids some of that medium's expressive imagery.

Rather than the idea that television is ontologically separate from the big screen, the resistance to style can be interpreted as a reactionary effort to prevent a stylistic heterogeneity. This counters the supposition, again borrowed from television history, that the medium's formal methods have become self-evident and codified through long use. The attempt to adapt *Day of the Triffids* (2009) can be seen as an opportunity to deploy not simply an excessive visuality, but a style that has been motivated by character development and thematic resonance. However, Caldwell makes another observation that:

> Television directing is no longer necessarily dominated by actor's directors. Veterans with decades of production experience, in fact, now argue that the best training for second-unit directing, a prerequisite for first-unit work, comes not from writing or producing, but from working as a cinematographer. You have to be a director of photography before you can direct second-unit, be-

> cause you have to be able to shoot in any style. As a director you have to match what the main cinematographer does. . . . Requiring a cinematography pedigree for television directing, could not be more alien to the script and acting sensitivity celebrated by many earlier television directors. John Frankenheimer and Delbert Mann in the 1950s could not have cared less about the grade of fog filter used on the set or the characteristic curves of their prime-time kinescope stocks.[51]

From this, it can be inferred that the danger to television drama is that CGI effects can be deployed to realize a poorly constructed narrative. Yet, the adaptation of a well-established novel such as *The Day of the Triffids*, as well as the reference to and possible homage to earlier adaptations, allows for a changed focus without interrupting the notion of good storycraft derived from a canonical text. The new adaptation avoided the earlier beginning of Masen waking up in hospital and the silence that Wyndham had described as "When a day that you happen to know is Wednesday starts off by sounding like Sunday, there is something seriously wrong somewhere." Instead, during the opening, there is a declaration by Masen that since the death of his mother when he was a child the problem presented to him of the Triffids is personal. The resulting action is much swifter and acts as a counterpoint to the original, as it leads to the discovery of an intruder in the Triffid farm, when Mason is stung by a Triffid and finally rushed to hospital in a frenetic attempt to save him. But the use of CGI remains restrained: this is clearly still a character-driven drama.

Nevertheless, Nick Copus, the director, was to admit that he intended the adaptation of his *Day of the Triffids* to be a return to "the 1980s and testosterone-fuelled flicks such as *Die Hard* for inspiration." Corpus explains:

> In 80s movies there's a lot of backlighting and use of smoke; things that aren't necessarily visually correct but that make shots more interesting. . . . Today people would say, "Why is there smoke in the shot?" They didn't care about that in the 80s and we don't care about that in this film. We've used so much smoke and dry ice. One of our mandates was not to have a really standard British drama look. [We wanted] to try and take it into a world that was more visually interesting and to give it that entertainment value, so we didn't fall too much into the doom and gloom of a film like *28 Days Later*.[52]

The main set piece of the show is at the end of episode one and builds into a cliff-hanger for the audience. Masen and Wilfred Coker, an American, have been captured by Torrance and driven to a deserted spot in the woods where the plan is to dispose of the two prisoners to the Triffids. The scene is shot using lighting to create the appropriately eerie atmosphere and sense of anticipation of something lurking in the shadows. Such carefully controlled lighting is a key signifier of how far television had travelled from the small screen and the aesthetics associated with it. The beams of the torches held by Torrance's men shine through a fine mist to add to the feelings of unearthliness as they drive deeper into the woods. Meanwhile, the sound of the Triffids is heard with mounting proximity. After allowing the dramatic tension to reach breaking point, the appearance of the Triffids is glimpsed incrementally rather than as a sudden revelation. Monstrous tentacles attack the men, blinding one, and allowing the others to hear his screams as he is extracted from the group and transported away to be consumed by the carnivorous plants. At last, by combining CGI and the actors in a complex, composite shot, the characters are confronted by the Triffids.

The second major confrontation between the Triffids and the remaining humans is at the end of episode two, again forming a set piece, which because of the possibility of home video can be watched and watched again. The spectacle can be enjoyed more thoroughly than when a program such as the original BBC adaptation of *Triffids* would be transmitted once and not seen again until it was repeated on broadcast television. The revolution in viewing television is beyond the scope of this book, but it can be said that the commodification of programs has increased apace since the opening of new markets using the technologies of video, satellite, and cable. All have represented an existential threat to older methods of broadcasting and consuming television because of the opportunity they afford of the repeated viewing of a single program. In many ways, television specificity ceases to be what it had been since the 1950s. By 2009, it had been transformed into a set of production conditions that sought not the ephemeral, but a distinctive product with lasting value. The aspiration to cinema of *The Day of the Triffids* is a sign of how much television has embraced digital imagery, in order to offer the epic scenes that once were the preserve of the large screen.

Masen, Jo, and the two girls, Susan and Imogen, are taken prisoner at their house, Shirning, once owned by Masen's father. Held prisoner by Torrance and his men, they plan to escape while the Triffids attack the house. The scene is set for a high action, slick production, and the use of CGI to seamlessly enhance the danger and violence levels of the program. However, despite the use of CGI, the story makes use of Masen's revelatory comprehension about how to avoid the stings of the Triffids. The realist disposition of television, if called into question by the digital environment and the danger of nonnarrative effects, is avoided when human performance once more becomes the central controlling agency for realizing the story.

The memory of the fate of his mother, who died helping to save her young son from the Triffids, is combined with a sudden realization of the teasing clues from Masen's past to suggest the correct action to protect them against the Triffids. Masen orders the others to give him time to think, as the reliance on human ingenuity rather than startling action completes the narrative. They discover that by smearing the poison from the Triffids onto their eyes it temporarily causes pain, but makes them undetectable to the mon-

Figure 5.1. Bill Masen demonstrates other ways of seeing the world.

strous plants. The plot revelation is therefore an intensification of television's intimacy: the limited field of human vision is restricted to the eyes of our characters. The investment in the characters is rewarded by this ingenious turn of events, making it possible to accept that there is an emotional core within the story rather than an endless succession of CGI effects. At some level, the 2009 adaptation demonstrates how television continues to resist the overreliance of CGI by not always drawing attention to its visuality to the detriment of a character-driven narrative.

CONCLUSION

The adaptation of *The Day of the Triffids* is a continuing discourse about the nature of the television image as its definitions have become unstable in an era dominated by digital technology. There has been both an acceptance as well as a resistance to a growing visibility within science fiction, which recalls earlier debates about the status of the genre on television. If spectacle can be made to fit into narrative, it is also true to say that science fiction has often exploited spectacle to make it part of the storytelling process. Although narrative and spectacle are not mutually exclusive, judgments about the aesthetic worth of a genre, which is often characterized by its ideas, appears to be at odds with visual display. For the science fiction genre to be accepted on television, it has sought to achieve a balance of plot and characterization and the avoidance of an excess of spectacle. Such avoidance also marks further qualitative differences between television and the cinema.

As we have seen, the initial television adaptation of *The Day of the Triffids* was the most faithful to the book, even more so than the 1968 radio adaptation. The least faithful was the 1962 film produced by MGM, which attracts almost universal derision as an example of what happens to a "good" book when adapted for the cinema. Notwithstanding the 1962 film, the aspiration to add *Triffids* to the canon is perhaps most noticeable as British television itself adapts to the changing media ecology of the digital age. *Triffids* as an example of postapocalyptic science fiction, which has been an enduring feature of British television science fiction, was expanded and developed in 2009 to engage with social issues and ideological concerns within the context of a mythology about the near future. Such issues address worries about environmental changes, including global warming,

In 2014, the announcement was made that *The Day of the Triffids* will receive the Hollywood treatment again. The intention of bringing back the story to the big screen highlights how the sheer number of adaptations of the same story has been criticized as Hollywood running out of original ideas. In this way, the difference between the last television adaptation and any future one at the cinema should be understood as the negative reliance on repeating conventions rather than offering a process that adds new canonical material to the world of the Triffids. The idea of canon, that certain things are true for an imaginary world, also demonstrates the continuing importance of the presence of an author. Wyndham operates as the originator, but today adaptation is about the franchising of the fictional item, though, compared to *Lord of the Rings*, for example, Wyndham's Triffids have contributed far less to the ongoing creation of a successful imaginary world. It is this lack that delimits the cultural impact of *Triffids*, while drawing attention to how its adaptation has been different from other science fiction texts, including *Doctor Who*. Such shows have spawned a host of ancillary texts and products to contribute to the sense of an unfolding narrative outside of the original book or program. The word *Triffids* may have long ago entered the generic lexicon, but, unlike other examples within the science fiction genre, *The Day of the Triffids* remains resistant to its adaptation beyond television, radio, or cinema, and to the offer of other possibilities, either philosophical or entertainment, outside the primary text. Whether this will change is a debate about how audiences are made aware of and how they can increase their understanding of a primary text. In the case of *Triffids*, the canon was reestablished by the 2009 adaptation, but, so far, it has avoided the addition of anything new.

NOTES

1. Aldiss, *Trillion Year Spree*, 315.

2. *Tonight Show*, originally broadcast September 6, 1960.
3. Aldiss, *Trillion Year Spree*, 313.
4. Polak, *The Image of the Future*, 183–219.
5. Lewis, "On Science Fiction," 109.
6. Wyndham, *Day of the Triffids*, 86.
7. Ibid., 151.
8. Ibid., 28.
9. *Hancock's Half Hour*, "Sunday Afternoon at Home," originally broadcast 1959.
10. Forrest J. Ackerman, "Book Reviews," *Astounding Science Fiction* (August 1951): 142.
11. Groff Conklin, "Five Star Shelf," *Galaxy Science Fiction* (August 1951): 99.
12. "Plant Power," *Radio Times*, September 5–11, 1981: 76.
13. Ibid.
14. Still frame, vari-speed, and picture during shuttle were all features not available on quad recorders without additional equipment.
15. Kimbley, "Survival of Celluloid," 22.
16. Jones, *Sony Guide to Audio and Video Recording*, 77–78.
17. Hall, "Television and Culture," 250. Hall is discussing the apparent transparency of the television studio.
18. Ibid.
19. Cellan-Jones, "Is There a Future," 19.
20. Kimbley, "Survival of Celluloid," 22.
21. Sexton, "Televerite Hits Britain," 431.
22. See Johnson, *Telefantasy*.
23. Gareth Preston, "Cult Television: The BBC's *Day of the Triffids*," http://www.westlakefilms.co.uk/pfculttriffids.htm.
24. John Walker quoted in Kimbley, "Survival of Celluloid," 21.
25. Derek Granger quoted in ibid., 22.
26. In 2014 and using the site Measuring Worth, http://www.measuringworth.com/.
27. Ellis, "Importance, Significance, Cost and Value," 49.
28. See Max Sexton, "The Origin of Gritty Realism."
29. Verity Lambert in Alvaredo and Stewart, *Made for Television*, 85.
30. Christie, "A Beginner's Guide," 10.
31. Ted Childs in Alvaredo and Stewart, *Made for Television*, 87.
32. Kuhn, *Alien Zone*, 9.
33. Diana Frampton, "Return of the Tripods to Exterminate the Daleks?" *Broadcast*, November 15, 1985, 24.
34. Taves, *Romance of Adventure*, 4–5.
35. Frampton, "Return of the Tripods."
36. From promotional flyer on Paintbox, "Paintbox in the DPC," *Quantel* (Autumn 1989).
37. Memo from Richard Bates to Brian Wenham (redirected to Jonathan Powell), October 24, 1985, BBC WAC T62/309/1/.
38. Memo from Jonathan Powell to Richard Bates, November 19, 1985, BBC WAC T62/309/1. Bates was Producer, Drama Series and Serials. Powell was Head of Series/Serials, Drama, Television.
39. Nelson, *State of Play*, 57.
40. Matthew Bell, "Has It Got the Fear Factor?," *Broadcast*, December 11, 2009, 28.
41. Ibid., 29.
42. Ibid.
43. Ibid.
44. Caldwell, *Televisuality*, 5.
45. Ellis, *Visible Fictions*, 162.
46. *The Private Life of Plants*, first broadcast 1995.
47. LaValley, "Traditions of Trickery," 144.
48. Bell, "Has It Got the Fear Factor?," 28–29.

49. Sam Wollaston, "The Day of the Triffids," *Guardian,* http://www.theguardian.com/tv-and-radio/2009/dec/30/day-of-the-triffids-review. ASBO is an acronym for "Anti-Social Behavior Order."
50. Black, *Reality Effect*, 211.
51. Caldwell, *Televisuality*, 77.
52. Bell, "Has It Got the Fear Factor?," 29.

Conclusion

The case study methodology adopted in this book precludes the drawing of sweeping conclusions and extrapolations. It is not simply that there is insufficient data to make an informed generalization, but rather that the basis of this approach, and the evidence presented here through it, are at odds with singular, overarching, ahistorical explanations. In arguing for a historically and culturally specific understanding of television, we see close attention to individual circumstances and contexts as the most appropriate way to understand the full complexity and richness of this period. Despite this, William Uricchio suggests, echoing earlier historians, "history's power derives from its ability to illuminate the present" and this study is no exception.[1]

Present-day debates about television, such as those of Uricchio and his fellow special-issue authors, center on the fate of this medium and other moving-image media in the light of digital technologies and especially the perceived convergence between media resulting from changing production, distribution, and exhibition practices and technologies. These include, but are not limited to, digital capture, storage, and editing; the shrinking of the theatrical window; on-demand streaming and Internet video; live cinema broadcasts; interactive video games; and the proliferation of screens in all spaces, whether domestic or public, including mobile and handheld screens. The study of the interaction between different media, including the crucial one between cinema and television, accentuated by the process of adaptation, might be expected to offer conclusions that engage straightforwardly with this widely accepted convergence account.

Consequently, the historical study of the period preceding the technological changes brought about by the "digital revolution" should provide evidence for clear medium specificities for television and cinema and the other media from which science fiction material was adapted. In such a reading, the essential qualities of the media are unavoidably technological, without human agency, and program makers adapt to it before the next technical development. Accordingly, such specificities are only eroded by the technological convergence of the digital. The opposite view, that the primacy of society constructs television, demands an understanding of the production culture surrounding it. Evidence for both approaches is found throughout this book, such as the difficulty expressed by the producers of *Out of the Unknown* of finding suitable material that fitted the medium of television. This had several factors beyond basic technological constraints peculiar to television. On the one hand, production limitations ruled out the epic scale of space exploration found in the source material; on the other hand, there were problems of adapting a story to the exigencies of an hour time slot within the weekly schedule. Moreover, the methods by which programs became commodities to be traded in the international market and presented as discrete items within the televisual flow played a key role. These multiple factors refute any singular explanation of sci-fi on television. Instead, the adaptation of science fiction for television is a complex discourse formed by television's institutional form, program formats (the series/anthology), economics, and audience.

The study of programs in this book has sought to theorize the adaptation of sci-fi for television as discourse, and, specifically, the genealogy of television form and the relations between media, even before digital technologies. Here medium specificity is found to be complicated not so much by the intrinsic nature of television, but because of its particular modes and the critical and institutional thinking behind its enormous output. This book has been continually alert to the question of television itself in order to deal satisfactorily with the issue of adaptation. When one talks about television, what sort of television is it? The earliest program makers of science fiction—Fred Ziv, Walt Disney, and Ivan Tors—realized that television had to be opened up and could never be limited to a particular type of program. Producers and the networks they worked for understood that they need not limit themselves to the single mode of live intimacy. Yet, the qualities of drama associated with live broadcasting were surrounded by uncritical mystique. At the heart of this mystique was the personalized intimacy, inherited from radio, between the broadcaster and the audience. For some practitioners at the time, the mechanical difficulties in the live presentation of immediate events simultaneously with their occurrence raised television's cultural status. But "live" transmission has been shown in this study to not be the only mode on television that possesses the dramatic power to reveal and inform.

The adaptation of science fiction offered multiple possibilities for the negotiation of medium specificity in the period examined here. First, the adaptation process naturally invited a consideration of the connections and distinctions between the media involved. Likewise the science fiction genre offered a space where experimentation in form was generically motivated and engagement with technology was not simply part of the story telling process but was embedded within many of the narratives as well, offering the opportunity to expand the aesthetic and technological definitions of what television could offer. The reworking of generic verisimilitude and its transformation in *Quatermass* and *Out of the Unknown* are cases in point.

The belief of many early practitioners that the cathode ray tube imposed limitations is evidenced in the attitudes of Rudolf Cartier to directing drama on television. Yet, although Cartier was excited by the possibility of the liveness of the small screen and its "hypnotic" effect on the individual watching it at home, he incorporated lighting effects that were less constrained than lighting for live broadcasts. Cartier's own formalist notions about the "essence" of film and television collapse in *Quatermass and the Pit*, which was able to utilize varied lighting because filming permitted frequent changes in players and objects in relation to light sources.

Other technical limitations informing the nature and form of the discourse about television adaptation are also revisited in this study. For instance, in this period television could not duplicate the complex editing seen in cinema, but the deployment of the vision mixer within the studio allowed for a series of dissolves and transitional fades in programs such as "No Place Like Earth" and "The Counterfeit Man" to produce complicated and deliberate effects. The cutting between cameras within the studio, controlled using the mixer in the control room, adapted editing techniques of repositioning the viewer within the narrative much as the cinema had done by cutting shots together. In this way, it was possible to adapt a literary script from John Wyndham and reproduce the introspection of his protagonist Ben Foster, in "No Place Like Earth." The use of allusion and symbolism to explore tensions and contradictions within complex ideas of civilization could be represented through the use of multiple visual and aural juxtapositions. Such possibilities, available by the mid-1960s, suggest avenues of experimentation that were opening up in television. These were not simply the direct adaptation of film techniques from the cinema, but attempts to find similitude as well as difference between competing methods of juxtaposing moving images and/or sound.

On television, as we have seen, the small screen was presumed to affect the tempo of camera movement and the need for the greater use of close-ups. However, this study demonstrates how narrative pacing was affected by modified techniques of editing the sound and images in television. The tempo of the montage sequences in "The Counterfeit Man" is extremely fast in places and throws doubt on the assertion that early television drama is "slow." The discontinuous use of space and time in the montage sequences was hybridized from film *and* techniques developed inside the "live" television studio. Such hybridity is striking because it existed not necessarily in a single program but between programs in the serial form of the anthology. In these ways, *Out of the Unknown* adapted, as well as developed, a diversity of modes for narrative progression and character subjectivity.

The incorporation of filmed material into live programs allowed practitioners to reflect again on the comparison between television and cinema. Like the cinema, the television camera can record the whole range of visual phenomena. However, if the live medium appears to have less capacity for fantasy and illusion, partly because it cannot modify the sense of time or space possible with film, Disney's "Man in Space" was able to engage its audience using several modes of expression. Using animation, real-life action, the appearance of Ward Kimball, and the German scientists commenting on the space race, it was possible to produce a program of selected recorded material and a script that suggested actuality and realism as well as fictional entertainment. Similarly, *Science Fiction Theatre* accommodated discussions about science ranging from the discursive anecdote to full-length fiction. In this way, the television program is both derivative of and different from film: The fidelity to real life goes back to the birth of cinema, but it has been remediated in television as immediacy at a historical moment when the new medium was seeking to experiment with new dramatic forms and modes.

In Britain, Nigel Kneale was to demonstrate that, if television's field was too narrow because of its heavy interest in the domestic, it was still possible to produce stories using a filmic mode that incorporated a set of ideas not yet seen on the new medium. We have seen how this was achieved in *Quatermass and the Pit*, a show that also incorporated a didactic speech at the end of the program to elucidate the purpose of the series. Again, the art of adaptation is revealed to lie in integrating new modes of drama and personal communication into the formal pattern of television narrative.

Crucially, such trends and developments on television have not been ignored in this study. The analysis of specific television programs serves to highlight major trends such as the introduction of telefilming in the United States and Britain or, at other times, local instances of innovation within the BBC, including thinking inside the Langham Group, although it was eventually regarded as a creative cul-de-sac. Rather than seeking monocausal determinism, the study seeks insight into how a series of large and small innovations enabled thematic expansion in the programs being examined. Such innovations were used by contemporary practitioners to explore the formal possibilities of television, but the book also acknowledges the resistance to adapting such possibilities because of the continuing need to appeal to a mass audience sitting at home rather than a specific audience interested in science fiction. The attitude of Leonard White, the producer of *Armchair Theatre*, is instructive here: he was averse to experimentation, but sought innovation and difference to avoid accusations of elitism.

As well as understanding how television communicates, examining what it communicates is revealing too. Science fiction's iconography of aliens and spaceships offers a break from the quotidian on television. However, such generic verisimilitude runs the risk of becoming mediocre. Again Kneale sought to adapt what he perceived to be the familiar and hackneyed in science fiction—spaceships and ray guns—to produce drama that could stand up to serious qualitative consideration. The gap between "serious" and "popular" drama is one that the study acknowledges and, to some degree, seeks to demonstrate because of the need to link differing modes to the range and scope of programs on television. If science fiction on television guaranteed familiar pleasures for the audience, the diversity of styles in *Out of the Unknown* was to offer elements of unexpectedness that could also compliment the notion of television as discovering new, original sources for adaptation, as well as providing an appeal to a mass audience. The same can be said of many of the other shows within this study, including the later programs *The Day of the Triffids* and *The Tripods*.

Although outside the scope of this book, this study inevitably raises important questions about television as a node within mass communications and its precise relationship to other types of communication technology. The cross-fertilization between several types of media technology seen here supports theoretical arguments against the concept of medium specificity. Other examples provided in this book for such an argument might be found in the role film technology played in the production of television shows, or the way a franchise such as *Superman* could be adapted for many different media with repeated success, indicating close links between them at both institutional and aesthetic levels. Yet *The New Adventures of Superman* is shown not simply to have eradicated medium specificity, but rather to have negotiated between existing medium definitions, using the liminality of science fiction and animation as a means of achieving this. It did not offer a simple endorsement of prevailing meanings of television or animation, but it was, nevertheless, embedded within those discourses. This is evident in both the style of the show and the stories it told, as well as discussions of the show in the trade and popular press.

This study demonstrates that medium specificity, despite the philosophical limitations and inconsistencies discussed in the introduction, did play an important role in the understanding of television in the period addressed here, primarily the 1950s through the 1980s. The choices made by producers were often informed by their understanding of what television was and what they felt could be achieved within the technological basis of those definitions. Similarly, viewers and critics used medium specificity as a key criterion for judging and understanding the television programs they watched. The study does not break with established ideas about the separate categories of cinema, radio, and television, so much as explore how types of media culture impose new interpretations on technologies that in themselves are unable to set the conditions of social change.

To return to the concerns that we began this conclusion with, how might these findings be used to consider present-day television productions? Our research, outlined in detail in the preceding chapters, does not offer a simple essentialized medium specificity for television, but neither should it be considered to indicate an undifferentiated continuation of convergence. Interactions of style or the collaborations between media and the routine use of adaptation within television indicate a degree of commonality between media that may be seen to presage the synergies and convergences of today's multimedia conglomerates. But our historical approach acknowledges that the changes that have occurred are, in part, visible because of established medium definitions, not least the established nomenclature that remains very much in evidence today: *television*, *radio*, and *cinema*. Thus the findings of this book, by rejecting "death-of-the-medium" accounts that depend upon singular definitions of media and technological determinism, may offer a model for considering the present-day impact of digital technologies while conversely also demonstrating the central role ideas of medium specificity have played in shaping and developing television.

Superman might again prove instructive in this regard. The lack of a singular originating source against which the 1960s animated television adaptation could be judged or verified and the way each adaptation of the character contributed to the overall canon can be seen as an early example of a transmedial franchise, a pattern that has become dominant in present-day moving-image production, with the Superman character, of course, remaining a commercially important example. Yet we have shown here how transmediality need not be interpreted as a flattening of the distinctions between media. The status of *Superman* as a transmedia narrative did not preclude the Filmation animated series or other adaptations continuing to engage with media definitions. On the contrary, having appeared in multiple media, the character offered a marker for differences between adaptations, most obviously in their variations in handling the balance between Clark Kent and Superman.

Similar arguments may be broached in relation to the technology of different media. The 2009 version of *The Day of the Triffids* adoption of CGI is evidence of the increasing availability to television of technologies associated with cinema and might thus be seen as a further example of convergence between them. Yet, as shown in chapter 5, this technology was used in a specific manner and was situated within a context that served to revive notions of televisuality rooted in intimacy, rather than simply ape cinematic spectacle. Technology does not dictate how it should be used, but it may come with expectations, which can either be accepted uncritically or navigated at an industrial or social level. While the notion of television as a singular, technologically determined medium has been shown to be both philosophically unsound and undermined by digital convergence, medium specificity remains a central discourse in discussion of moving images. As a result the approach adopted here becomes more, not less, valuable.

NOTE

1. Uricchio, "Film, Cinema, Television . . . Media?," 266.

Bibliography

Aldiss, Brian. *Trillion Year Spree: The History of Science Fiction*. London: Paladin Books, 1988.

Allen, Mike. *Live from the Moon: Film, Television and the Space Race*. London: I. B. Tauris, 2009.

Altman, Rick. "Television Sound." In *Studies in Entertainment: Critical Approaches to Mass Culture*, edited by Tania Modleski, 39–54. Bloomington: Indiana University Press, 1986.

Alvarado, Manuel, and John Stewart. *Made for Television, Euston Films Limited*. London: Thames Television International, 1985.

Anderson, Gary H., *Video Editing and Post-Production: A Professional Guide*. White Plains NY: Knowledge Industry Publications, 1988.

Annan, Lord. "The Relationship of Management with Creative Staff." *Report of the Committee on the Future of Broadcasting*. London: HMSO, 1977.

Backer, Ron. *Gripping Chapters: The Sound Movie Serial*. Albany, GA: BearManor Media, 2010.

Baily, Kenneth, ed. *The Television Annual for 1950/51*. London: Odhams Press, 1951.

Bakewell, Joan, and Nicholas Garnham. *The New Priesthood: British Television Today*. London: Allen Lane, 1970.

Ballard, J. G. "Which Way to Inner Space?" *New Worlds* 118 (May 1962): 2–3 and 116–18.

Barr, Charles. "They Think It's All Over: The Dramatic Legacy of Live Television." In *Big Picture, Small Screen: The Relations between Film and Television*, edited by John Hill and Martin McLoone, 47–75. Luton, UK: John Libbey Media, 1996.

Barrier, Michael. *Hollywood Cartoons: American Animation in Its Golden Age*. Oxford: Oxford University Press, 1999.

Barrier, Michael, and Bill Spicer. "An Interview with Chuck Jones." *Funnyworld* 13 (1971): 4–19.

Barry, Michael. *From the Palace to the Grove*. London: Royal Television Society, 1992.

Batchelor, David. *Chromophobia*. London: Reaktion, 2000.

BBC Handbook. London: The British Broadcasting Corporation, 1938.

BBC Handbook. London: The British Broadcasting Corporation, 1958.

Black, Joel. *The Reality Effect: Film Culture and the Graphic Imperative*. London: Routledge, 2002.

Bluestone, George. *Novels into Film*. Baltimore: Johns Hopkins University Press, 1957.

Boddy, William. *Fifties Television: The Industry and Its Critics*. Chicago: Illini Books, 1993.

———. "The Studios Move into Prime Time: Hollywood and the Television Industry in the 1950s." *Cinema Journal* 24, no. 4 (Summer 1985): 23–37.

Bordwell, David. "The Art Cinema as a Mode of Practice." *Film Criticism* 4, no. 1 (1979): 56–64.

Broughton, Irv. *Producers on Production: The Making of Film and Television*. Jefferson, NC: McFarland, 1986.

Buscombe, Edward. "All Bark and No Bite: The Film Industry's Response to Television." In *Popular Television in Britain: Studies in Cultural Theory*, edited by John Corner, 197–207. London: British Film Institute, 1991.

Bussell, Jan. *The Art of Television*. London: Faber and Faber, 1952.

Butcher, William. "Introduction." In Jules Verne, *Twenty Thousand Leagues under the Sea/Vingt mille lieues sous les mers*. Oxford: Oxford University Press, 1998.

Caldwell, John. *Televisuality: Style, Crisis, and Authority in American Television*. New Brunswick, NJ: Rutgers University Press, 1994.

Cardwell, Sarah. *Adaptation Revisited: Television and the Classic Novel*. Manchester, UK: Manchester University Press, 2002.

———. "Television amongst Friends: Medium, Art, Media." *Critical Studies in Television* 9, no. 3 (2014): 6–21.

Carroll, Noel. *Engaging the Moving Image*. New Haven, CT: Yale University Press, 2003.

————. "Medium Specificity Arguments and Self-Consciously Invented Arts: Film, Video and Photography." *Millenium Film Journal* 14/15 (1984): 127–153.

————. "The Specificity of Media in the Arts." *Journal of Aesthetic Education* 19, no. 4 (1985): 5–20.

————. *Theorizing the Moving Image*. Cambridge: Cambridge University Press, 1996.

Cartier, Rudolph. "A Foot in Both Camps." *Films and Filming*, September 1958, 10 and 31.

Castle, Terry. *The Female Thermometer: Eighteenth-Century Culture and the Invention of the Uncanny*. Oxford: Oxford University Press, 1995.

Caughie, John. "Adorno's Reproach: Repetition, Difference and Television Genre," *Screen* 32, no. 2 (Summer 1991): 127–53.

———. *Television Drama: Realism, Modernism and British Culture*. Oxford: Oxford University Press, 2000.

Cellan-Jones, James. "Is There a Future for the Single Play?" [Special Panel Discussion] *Television: Journal of the Royal Television Society* 16, no. 11 (September–October 1977): 19.

Chapman, James. "Quatermass and the Origins of British Television Science Fiction." In *British Science Fiction Television: Hitchhiker's Guide*, edited by John Cook and Peter Wright, 21–51. London: I. B. Tauris, 2006.

Christie, Ian. "A Beginner's Guide to the Telefilm Jungle." In *The Screen on the Tube: Filmed TV Drama*, edited by Jayne Pilling and Kingsley Canham, 10–11. Norwich: Cinema City, 1983.

Cook, John, and Peter Wright. *British Science Fiction Television: Hitchhiker's Guide*. London: I. B. Tauris, 2006.

Crafton, Donald. *Shadow of a Mouse: Performance, Belief, and World-Making in Animation*. Berkeley: University of California Press, 2013.

Crisell, Andrew. *An Introductory History of British Broadcasting*. London: Routledge, 1997.

Daniels, Les. *Superman: The Complete History*. San Francisco: Chronicle, 1998.

Ellis, John. "Importance, Significance, Cost and Value: Is a TV Canon Possible?" In *ITV Cultures: Independent Television Over Fifty Years*, edited by Catherine Johnson and Rob Turnock, 36–56. Maidenhead, UK: Open University Press, 2005.

———. "Is It Possible to Construct a Canon of Television Programmes?" In *Re-viewing Television History: Critical Issues in Television Historiography*, edited by Helen Wheatley, 15–26. London: I. B. Tauris, 2007.

———. *Visible Fictions: Cinema: Television: Video*, revised edition. London: Routledge, 1992.

Feuer, Jane. "The Concept of Live Television: Ontology as Ideology." In *Regarding Television*, edited by E. Kaplan, 12–22. Frederick, MD: University Publications of America, 1983.

———. "Genre Study and Television." In *Channels of Discourse: Television and Contemporary Criticism*, edited by Robert C. Allen, 113–33. London: Methuen, 1987.

Fleischer, Richard. *Out of the Inkwell: Max Fleischer and the Animation Revolution*. Lexington: University Press of Kentucky, 2005.

Forman, Denis. "Television in 1959." In *Granada: The First 25 Years*. [Dossier #9], edited by Edward Buscombe, 60. London: British Film Institute, 1981.

Gielgud, Val. "Policy and Problems of Broadcast Drama." *BBC Quarterly* 2, no. 1 (1947): 18–23.

Goldschmidt, Rick. *The Enchanted World of Rankin/Bass: A Portfolio*. Issaquah, WA: Tiger Mountain Press, 1997.

Gomery, Douglas. "Theatre Television," *Velvet Light Trap* 21 (1985): 54–61.

Grams, Martin, Jr. *Science Fiction Theatre: A History of the Television Program*. Albany, GA: BearManor Media, 2011.

Hall, Stuart. "Television and Culture." *Sight and Sound* 45, no. 4 (Autumn 1976): 246–52.

Hewett, Richard. "Acting for Auntie: From Studio Realism to Location Realism in BBC Television Drama." Unpublished thesis, University of London, 2012.

Hill, John. "Creative in Its Own Right: The Langham Group and the Search for a New Television Drama." In *Experimental British Television*, edited by Laura Mulvey and Jamie Sexton, 17–30. Manchester, UK: Manchester University Press, 2007.

Hillman, Donald Stuart. "More Than Meets the Eye." In *How to Direct for Television*, 75–79. New York: Hastings House, 1955.

Honri, Baynham. "Make Sure We Sell Our Share of Packages in the TV Supermarket." *Production Review*, supplement to *Kinematograph Weekly*, November 5, 1964, 11.

Hume, Alan. "Filming for Colour Television Series." *British Kinematography Sound and Television* 50, no. 1 (January 1968): 4–8.

Jacobs, Jason. *The Intimate Screen: Early British Television Drama*. Oxford: Oxford University Press, 2000.

———. "Issues of Judgement and Value in Television Studies." *International Journal of Cultural Studies* 4, no. 4 (December 2001): 427–47.

Johnson, Catherine. *Telefantasy*. London: British Film Institute, 2005.

Jones, Mike. *The Sony Guide to Audio and Video Recording*. Slough, UK: Sony UK Limited, 1982.

Kennedy Martin, Troy. "Nats Go Home: First Statement of a New Drama for Television." *Encore* 48 (March/April, 1964): 21–33.

———. "Sharpening the Edge of TV Drama" (an edited version of the MacTaggart lecture). *Listener*, 28 August 1986: 12.

———. "*Up the Junction* and After." *Contrast* 4, nos. 5–6 (Winter 1965/Spring 1966): 139–40.

Kimbley, D. J. "Survival of Celluloid in the Age of Microelectronics." *Television: the Journal of the Royal Television Society*, September/October 1981, 21–25.

Kneale, Nigel. "Not Quite So Intimate." *Sight and Sound* 28, no. 2 (Spring 1959): 86–88.

Kohl, Leonard. "Flash Back to the Future: From Comic Strip to Serial Sci-Fi." *Filmfax* 45 (June/July 1994): 46–58.

Kotcheff, Ted. "The Making of Life at the Top." *Film and Television Technician* 31 (December 1965): 230–32.

Kuhn, Annette, ed. *Alien Zone: Cultural Theory and Contemporary Science Fiction Cinema*. London: Verso, 1990.

Lafferty, William."Television." In *Film and the Arts in Symbiosis*, edited by Gary R. Edgerton, 273–309. New York: Greenwood Press, 1988.

LaValley, Albert J. "Traditions of Trickery: The Role of Special Effects in the Science Fiction Film." In *Shadows of the Magic Lamp*, edited by George Slusser and Eric S. Rabkin, 141–58. Carbondale: Southern Illinois University Press, 1985.

Leitch, Thomas M. *Film Adaptation and Its Discontents: From "Gone with the Wind" to "The Passion of the Christ."* Baltimore: Johns Hopkins University Press, 2007.

Leman, Joy. "Wise Scientists and Female Androids: Class and Gender in Science Fiction." In *Popular Television in Britain: Studies in Cultural History*, edited by John Corner, 108–24. London: British Film Institute, 1991.

Lewis, C. S. "On Science Fiction." In *Science Fiction*, edited by Mark Rose, 103–15. Englewood Cliffs, NJ: Prentice-Hall, 1976.

Luciano, Patrick, and Gary Colville. *American Science Fiction Television Series of the 1950s*. Jefferson, NC: McFarland, 1998.

Mackie, Philip. "Six Hundred Hours a Week." *Sight and Sound* 24, no. 1 (July–September 1954): 45–48.

Madden, Cecil. "Television: Problems and Possibilities." *BBC Quarterly* 2, no. 4 (1948): 225–28.

Maltin, Leonard. *The Disney Films*. 4th ed. New York: Disney Editions, 2000.

———. *Of Mice and Magic: A History of American Animated Cartoons*. New York: Plume, 1980.

Mittel, Jason. "The Great Saturday Morning Exile: Scheduling Cartoons on Television's Periphery in the 1960s." In *Prime Time Animation: Television Animation and American Culture*, edited by Carol Stabile and Mark Harrison, 33–54. Abingdon, UK: Routledge, 2003.

Morgan, T. J. *The True Book about Television and Radio*, revised edition. London: Frederick Muller, 1961.

Mulvey, Laura. *Death 24x a Second: Stillness and the Moving Image*. London: Reaktion, 2006.

Murdock, Graham. "Authorship and Organisation." *Screen Education* 35 (Summer 1980): 19–34.

Myles, Lynda, and Julian Petley. "Rudolph Cartier." *Sight and Sound* 59, no. 2 (Spring 1990): 126–29.

Nelson, Robin. *State of Play: Contemporary High-End TV Drama*. Manchester, UK: Manchester University Press, 2007.

Pattison, Barrie. "The Man to Beat: Rudolph Cartier." *Film* 64 (Winter 1971): 17–18.

Perlmutter, David. *America Toons In: A History of Television Animation*. Jefferson, NC: McFarland, 2014.

Petley, Julian. "The Manxman: The Career of Nigel Kneale." *Monthly Film Bulletin* 56, no. 662 (March 1989): 90–93 and 95.

Pixley, Andrew. "The Martian Inheritance: We Are the Martians . . ." *TV Zone* 110: 36–41.

Polak, Fred. *The Image of the Future*. Amsterdam: Elsevier, 1973.

Porter, Vincent, and Chris Wicking. "The Making of *Life at the Top*." *Film and Television Technician* 31 (December 1965): 230–32.

Porterfield, John, and Kay Reynolds, eds. *We Present Television*. New York: Norton, 1940.

Purser, Philip. "Landscape of TV Drama." *Contrast* 1, no. 1 (Autumn 1961): 11–20.

Roberts, Edward Barry. "Writing for Television." In *The Best Television Plays 1950–51*, edited by William Kaufman, 295–300. New York: Merlin Press, 1952.

Salent, Richard S., Thomas K. Fisher, and Leon R. Brookes. "The Functions and Practices of a Television Network." In *Law and Contemporary Problems* 22 (Autumn 1957): 584–611.

Sandbrook, Dominic. *Never Had It So Good: A History of Britain from Suez to the Beatles*. London: Little Brown, 2005.

Schatz, Thomas. *Hollywood Film Genre as Ritual: A Theoretical and Methodological Inquiry*. Iowa City: University of Iowa Press, 1976.

———. "The Structural Influence: New Directions in Film Genre Study." *Quarterly Review of Film Studies* 2, no. 3 (1977): 301–12.

Scheimer, Lou. *Lou Scheimer: Creating the Filmation Generation*. Raleigh, NC: TwoMorrows, 2012.

Scivally, Bruce. *Superman on Film, Television, Radio, and Broadway*. Jefferson, NC: McFarland, 2008.

Sexton, Jamie. "Televerite Hits Britain: Documentary, Drama and the Growth of 16mm Filmmaking in British Television." *Screen* 44, no. 4 (Winter 2003): 429–44.

Sexton, Max. "The Origin of Gritty Realism: Euston Films and *Special Branch*." *Journal of British Cinema and Television* 11, no. 1 (January 2014): 23–40.

Shubik, Irene. *Play for Today: the Evolution of Television Drama*. Manchester, UK: Manchester University Press, 2000.

———. "Producers and Directors." In *The New Priesthood: British Television Today*, edited by Joan Bakewell and Nicholas Garnham, 87–89. London: Allen Lane, 1970.

Smith, Bernard B. "Television—There Ought to Be a Law." *Harper's* 197, no. 1180 (September 1948): 34–42.

Sontag, Susan. *Against Interpretation*. New York: Dell, 1966.

Stableford, Brian, *Scientific Romance in Britain 1980–1950*. London: Fourth Estate, 1985.

Stasheff, Edward, Rudy Bretz, John Gartley, and Lynn Gartley. *The Television Program: Its Direction and Production*. New York: Hill and Wang, 1976.

Taves, Brian. "The Making of *20,000 Leagues under the Sea*." *Filmfax* 58 (1996): 44–54, 138–39.

———. *The Romance of Adventure: The Genre of Historical Adventure Movies*. Jackson: University Press of Mississippi, 1993.

Taylor, Don, *Days of Vision—Working with David Mercer: Television Drama Then and Now*. London: Methuen, 1990.

Telotte, J. P. "The Cinematic Zone of *The Twilight Zone*." In *Science Fiction Film, Television, and Adaptation: Across the Screens*, edited by J. P. Telotte and Gerald Duchovnay, 20–34. Abingdon, UK: Routledge, 2012.

———. "Man and Superman: The Fleischer Studio Negotiates the Real." *Quarterly Review of Film and Video* 27, no. 4 (2010): 290–98.

———. *The Mouse Machine: Disney and Technology*. Urbana: University of Illinois Press, 2008.

———. "Science Fiction as 'True-Life Adventure': Disney and the Case of *20,000 Leagues under the Sea*." *Film & History* 40, no. 2 (2010): 66–79.

Telotte, J. P., and Gerald Duchovnay, eds. *Science Fiction Film, Television, and Adaptation: Across the Screens*. Abingdon, UK: Routledge, 2012.

Thomas, Frank, and Ollie Johnston. *The Illusion of Life: Disney Animation*. New York: Disney Editions, 1984.

Todorov, Tzvetan. *The Fantastic: A Structural Approach to a Literary Genre*. Ithaca, NY: Cornell University Press, 1975.

Uricchio, William. "Film, Cinema, Television . . . Media?" *New Review of Film and Television Studies* 12, no. 3 (2014): 266–79.

Vahimagi, Tise. "Fantastic Cliffhangers." *Starburst* 2, no. 2 (1979): 34–38.

Verne, Jules. *Twenty Thousand Leagues under the Sea/Vingt mille lieues sous les mers*. Translated by William Butcher. Oxford: Oxford University Press, 1998.

———. *Twenty Thousand Leagues under the Seas*. Translated by Unknown. Boston: George M. Smith, 1875.

———. *Twenty Thousand Leagues under the Seas*. Translated by Unknown. London: Ward, Lock & Co., 1892.

Ward, Mark. *Out of the Unknown: A Guide to the Legendary BBC Series*. Bristol, UK: Kaleidoscope, 2004.

Watts, Steven. *The Magic Kingdom: Walt Disney and the American Way of Life*. Columbia: University of Missouri Press, 1997.

Western, John. "Television Girds for Battle." In *Television Broadcasting: Production, Economics, Technique*, edited by Lenox Lohr, 547–63. New York: McGraw-Hill, 1940.

Wheatley, Helen, "And Now for Your Sunday Night Experimental Drama . . . : Experimentation and *Armchair Theatre*." In *Experimental British Television*, edited by Laura Mulvey and Jamie Sexton, 31–47. Manchester, UK: Manchester University Press, 2007.

Wheen, Francis. *Television: The Book of the Major ITV Series*. London: Century Publishing, 1985.

Wheldon, Huw. "BBC Executives." In *The New Priesthood: British Television Today*, edited by Joan Bakewell and Nicholas Garnham, 223–25. London: Allen Lane, 1970.

Williams, Raymond. *Television: Technology and Cultural Form*. London: Fontana/Collins, 1974.

Wilson, Donald. In "Introduction." Michael Barry, *The Television Playwright*. London: Michael Joseph, 1960.

Wyndham, John, *The Day of the Triffids*. London: Penguin, 1972.

Zubatkin, Marc. "Strock Footage: An Interview with Herbert L. Strock." *Filmfax*, no. 55 (March/April 1996): 34–41.

Index

About the Authors

Max Sexton is a lecturer at the University of Surrey, having previously worked at the University of London, where he currently teaches film theory and British cinema. After working in British television, he ran his own production company for several years, and worked as a journalist, before completing an MA in History of Film and Visual Media (with distinction) and a PhD at Birkbeck, University of London. His PhD thesis was on the specificity of 16 mm and 35 mm film on television as an example of a production and textual strategy in Britain and America. He is currently interested in adaptation and other intermedial areas between film and television, including debates about the use of live performance and special effects. He is also researching how the performance, presentation, and representation of magic and illusion can suggest new ways of thinking about the relations between film and television.

Malcolm Cook is a lecturer at Middlesex University and Central Saint Martins, University of the Arts, London. He was awarded a PhD at Birkbeck, University of London in 2013 for his doctoral thesis "Animating Perception: British Cartoons from Music Hall to Cinema, 1880–1928," which addressed early British animated cartoons prior to the advent of sound cinema, with a particular focus on the relationship between the moving image and the graphic arts and other pre-cinematic entertainments. He has published a number of chapters and articles on animation, early cinema, and their intermedial relationships.

Lightning Source UK Ltd.
Milton Keynes UK
UKOW02n2053040815

256367UK00001B/20/P

9 781442 252691